Business Guides on the Go

"Business Guides on the Go" presents cutting-edge insights from practice on particular topics within the fields of business, management, and finance. Written by practitioners and experts in a concise and accessible form the series provides professionals with a general understanding and a first practical approach to latest developments in business strategy, leadership, operations, HR management, innovation and technology management, marketing or digitalization. Students of business administration or management will also benefit from these practical guides for their future occupation/careers.

These Guides suit the needs of today's fast reader.

Patricia A. Adam

Agile in ISO 9001

How to Integrate Agile Processes into Your Quality Management System

 Springer

Patricia A. Adam (iD)
Hochschule Hannover
Hannover, Germany

ISSN 2731-4758   ISSN 2731-4766   (electronic)
Business Guides on the Go
ISBN 978-3-031-23587-0   ISBN 978-3-031-23588-7   (eBook)
https://doi.org/10.1007/978-3-031-23588-7

Für alle, die Elefanten das Fliegen beibringen

For all that teach elephants how to fly

Preface

Agility is "in fashion". If you enter the keyword "Agility" at Amazon and search for specialised business books in English, you will receive more than 1000 suggestions. If you choose the keyword "agile", the number even increases to more than 10,000 results (Amazon, 2022). Judging by the titles—and usually the contents—of these books, agility is "the" solution to almost all current management problems: Agile organisations are supposedly more dynamic and flexible (Scheller, 2017). You react faster to changes and can be leaner (Foegen & Kaczmarek, 2016). If you lead in an agile way, you can achieve higher performance and creativity in teams (Hofert, 2016) and thus successfully bring your organisation into the digital future (Hübler, 2018).

The aim of this book is to allow a realistic assessment of agility. It should not only define agility and distinguish it from flexibility. It is also intended to provide you with guidelines for the opportunities and limitations of using this modern variation of organisational design in certified quality management systems based on ISO 9001.

Universally, agility is considered to be a remedy against all kinds of issues that can affect an organisation. Thus, a modern organisation is at least expected to experiment with agile approaches and, for example, set up "agility labs—workspaces that feature walls to write on, colourful beanbags, and tables on wheels. Whoever does not subscribe to this view of agility increasingly gets the reputation of being unprogressive and not future-proof. Within the organisations, this fate particularly affects quality and process management departments. They traditionally rely on controlled processes, documented procedures, and verification of specifications. Because of their mandate and organisational role, they have difficulty dealing with nebulous promises and obscure conditions, especially if they are responsible for maintaining a management system that is certified based on ISO 9001.

It is all too understandable that the promises agility holds inspire managers which strive to transfer their enthusiasm quickly to their employees and organisations. However, notions of agility often lack substance and a realistic assessment of their potential contribution to organisational development. Agility fans on the one hand ("everything must be agile") and process fans on the other ("processes are controlled—agility is just chaos") are easily irreconcilably opposed to each other. This might lead to devastating consequences for the organisational cooperation.

This book is the result of my research project "Certification of agile processes". For this purpose, I first conducted in-depth interviews with representatives of companies from various industries who already had practical experience with the use of agile approaches. In particular, I asked about their understanding of agility, typical examples for the use of agile approaches, the results and challenges that have arisen, and any experiences in audit situations. The knowledge gained was supplemented by a comprehensive literature analysis and experiences from my personal auditing practice. The resulting view on agility was in turn transferred to ISO 9001:2015.

This much in advance: Agility is neither the legendary Swiss Army knife nor the sheer chaos of unplanned activities. Instead, it is an important addition to the process- and project-oriented forms of organisational development that have dominated until recently. So, agility is worth discovering!

Hannover, Germany   Patricia Adam

References

Amazon. (2022). Results or suggestions for books: "agility" or "agile". Amazon. Accessed September 21, 2022, from https://www.amazon.de/s?k=agile&rh=n%3A288100&__mk_de_DE=%C3%85M%C3%85%C5%BD%C3%95%C3%91&ref=nb_sb_noss

Foegen, J. M., & Kaczmarek, C. (2016). *Organisation in einer digitalen Zeit. Ein Buch für die Gestaltung von reaktionsfähigen und schlanken Organisationen mit Hilfe von Scaled Agile & Lean Mustern. Dritte Auflage.* Wibas GmbH.

Hofert, S. (2016). *Agiler führen.* Springer Fachmedien Wiesbaden.

Hübler, M. (2018). *Menschlich - Demokratisch - Agil. Agilität mit Verantwortung. 1. Auflage.* Walhalla und Praetoria (Metropolitan Bücher).

Scheller, T. (2017). *Auf dem Weg zur agilen Organisation. Wie Sie Ihr Unternehmen dynamischer, flexibler und leistungsfähiger gestalten. 1. Auflage.* Verlag Franz Vahlen.

References

About the Book

What you will find in this book:

- A clear definition and differentiation of agility and agile practices as an addition to processes and projects
- The opportunities and limitations of using agile practices in ISO 9001-certified management systems
- Typical obstacles during the reconciliation of agile practices with the "classic" organisational environment
- Various practical examples from different industries and departments

This information will be provided in the following order:

In the first chapter, the basic concepts of agility, agile practices, agile teams, as well as agile procedures and methods are defined and clearly distinguished from the classical way of organising activities. In addition, there are recommendations as to when the use of agile practices is worthwhile and which agile methods are frequently used.

In the second chapter, specific examples are used to explain how agile practices can be integrated into a management system according to ISO 9001:2015. Particular emphasis is placed on the assumption of planning and monitoring activities by self-controlling agile teams, the control of agile processes with the help of the so-called BIG FIVE, and the creative documentation of agile practices.

The third chapter is dedicated to typical stumbling blocks occurring during the integration of agile practices in everyday business. The design of fitting interfaces to classical (sub-) processes, the selection of suitable agile team members, the handling of redundant managers, and the challenges of retaining common Human Resource Management (HRM) instruments are the biggest obstacles for a successful integration.

The fourth and final chapter deals with the rise of agile practices in the disruptive environment of the "new normal". It explains why a broad introduction of home office forces organisations into agility. In this situation, the use of agile tools and a reliable output measurement provides much needed support. The chapter closes with an outlook of the introduction of the concept of agility into the EFQM model as well as into ISO management standards.

Contents

About the Author

Dr. Patricia A. Adam is Professor of International Management at Hochschule Hannover, University of Applied Sciences and Arts. She is an expert in organisational development and deals with management systems as well as intercultural issues. Patricia Adam is a member of the German Association for Quality (DGQ) and the German Speakers Association (GSA). As part-time ISO 9001 auditor and EFQM assessor, she regularly travels the world in order to assess organisations and their management systems. This book is a result of a research project funded by the DGQ.

List of Abbreviations

CPE	Continuing Professional Education
DGQ	Deutsche Gesellschaft für Qualität e.V. (German Association for Quality)
EFQM	Original: European Foundation for Quality Management
HRM	Human Resource Management
ICS	Internal Control System
QM	Quality management
QMS	Quality management system
VUCA (world)	Volatile—Uncertain—Complex—Ambiguous (world)

List of Figures

List of Figures

1

Agile in Organisations: What Is That?

Abstract In this chapter, the basic concepts of agility, agile practices, agile teams, and agile procedures and methods are defined and clearly distinguished from the classical way of organising activities. In addition, there are recommendations as to when the use of agile practices is worthwhile and which agile methods are frequently used.

Keywords Agility • Agile practices • VUCA • Agile teams • Quality management • Agile methods • Project • Process • Agile process • Organisational design

1.1 Agility, Agile Practices, and the VUCA World

Agility in business is quite in fashion. However, the understanding of agility is as broad as the agile approaches used. A common understanding of what agility exactly means is hardly existing—neither in practice nor in science. The understanding of agility in practice is particularly divergent. Within the scope of in-depth interviews conducted for the research project "Certification of Agile Processes" there was a lot of variety. The definition of agility ranged from individual terms such as "self-control", "flexible reaction to change", and the general "dealing with the unplanned" to precise tools: "Agility is transition to Scrum". As companies became more concerned with agility, the understanding became more complex, so that terms such as "mindset", "values", "process models", and "methods" started to be used more frequently. Often, agility was also defined by what it is not: "Agile does not mean haphazard". Thus, agility and agile practices were clearly distinguished from an everyday kind of flexibility which is often just "laissez-faire" or "management by chaos". Occasionally, agility was understood as a fundamental counter-model to a process-related approach. The multitude of (specialist) publications offer little help, as these publications are also far from a uniform definition. What exactly constitutes an agile approach? Characteristics that clearly distinguish agility from previously common approaches are often not described at all. Only in very rare cases descriptions are specific. Thus, the overall view on characteristics of agility is very inconsistent. However, there is consensus in theory and practice that agility has something to do with flexibility and should help to manage activities in an increasingly uncertain and complex world.

It therefore makes sense to start from the common baseline: enabling flexibility to survive in the VUCA world. The acronym **VUCA** stands for a world in which exact planning is no longer possible because it is characterised by **v**olatility, **u**ncertainty, **c**omplexity, and **a**mbiguity, as shown in Fig. 1.1.[1]

- **Volatility:**
  Customer needs do change almost without notice and capital markets are subject to extreme fluctuations.
- **Uncertainty:**
  It is becoming increasingly clear that the future cannot be planned. This makes it all the more difficult to develop strategies for such an environment.
- **Complexity:**
  Dependencies in global supply chains are becoming increasingly difficult to manage. The same applies to IT: While the space shuttle still had 400,000 lines of code, Windows Vista 2007 already had 50 million lines of code, SAP has 500 million lines of code, and Google has

---

[1] A very detailed examination of the VUCA world can be found in Scheller (2017), pp. 19–42.

Fig. 1.1 The VUCA world (own illustration)

more than 2 billion lines of code (Information is Beautiful, 2015). It is therefore not surprising that the software industry is a pioneer in the use of new organisational concepts to master the VUCA world.

- **Ambiguity:**
  The ambiguity of a situation leads to the equivalence of different solutions for problems that have been dealt with. There is no longer the one perfect solution. Instead, there are several equally good (or bad) alternatives that must be evaluated and selected carefully.

The VUCA world causes considerable changes in markets, business models, and internal processes and thus forces organisations to adapt. An organisation[2] confronted with the VUCA world can no longer base its operations on controlled conditions. It is required to deal with more flexible forms of organisational structure and process design and to establish other forms of control. This is exactly what agility was developed for.

---

[2] "Organisation" is understood in the comprehensive sense of ISO 9000:2015, 3.2.1 and thus includes all persons and groups of persons who train their own functions to achieve goals: Sole proprietors, companies with or without a profit-making purpose, associations, clubs, authorities and institutions of all kinds, whether public or private, registered or not.

Note: Agility is considered a panacea against the VUCA world.

In contrast to numerous publications that can be reduced to the simple formula "Agile = new and hip, process management = old and stupid", a realistic view of the management of today's organisations shows that different business units or departments are subject to different conditions. In this respect, it makes sense not to be guided by a 100% model but by the idea of an appropriate degree of agility. This is also discussed in more detail in Sect. 1.5.

Accordingly, agility in the organisational context can be defined as follows:[3]

Definition: Agility in organisations

An agile organisation understands that dealing with constant uncertainty and the resulting unplanned situations is a natural part of its existence. Therefore, it systematically integrates these situations into the management of its activities. The degree of agility of an organisation is determined by the use of agile practices and the alignment with agile values and principles.

For practical use and differentiation from flexibility, it is necessary to add further information to this general definition. Analogous to the approach used in ISO 9000:2015, necessary additions are made by means of notes:

Notes on the definition of agility in organisations

Note 1: *Agile values and principles are based on the four Agile Values and twelve Agile Principles defined in the Agile Manifesto and are to be customised depending on the organisation, its environment and industry.*

The Agile Manifesto is the milestone and starting point of today's basic understanding of agility. It was developed in 2001 by Jim Highsmith and a group of agile software developers ("The Agile Alliance"), published

---

[3] Most of the following definitions of agility and agile practices were first published under Creative Commons in Adam 2018.

We value...

individuals and interactions over
processes and tools

working software over comprehensive documentation

customer collaboration over
contract negotiation

responding to change over following a plan

While there is value in the items on the right,

we value the items on the left more.

Fig. 1.2 Values of the Agile Manifesto (own illustration, text taken from The Agile Alliance, 2001)

openly on the Internet and has since been translated into 68 languages (The Agile Alliance, 2001). The 4 agile values and 12 agile principles defined therein essentially describe what is known in most recent publications as the "agile mindset" (see also Sect. 3.2). The Agile Manifesto argues in favour of a new set of guiding values for agile software development, shifting the emphasis to individuals and responsive interactions as depicted in Fig. 1.2.

In recent publications, it is often overlooked that the agile manifesto does not call for an abolishment of process-based thinking, documentation, contracts, and planning. It is rather about shifting the emphasis towards a more open-minded and responsive approach.

This requires organisations and their employees at all levels to have a certain attitude and the willingness to take responsibility for sustainable, self-organised organisational development. Therefore, the introduction of agility is often accompanied by a significant cultural change. Benedikt Sommerhoff has published an interpretation and further development of agile quality management based on the Agile Manifesto for the German Association for Quality (DGQ) (Sommerhoff, 2016).

For the examination of the possible use of agile practices in a quality management system (QMS), it is necessary to provide a precise definition and a clear differentiation of agile practices from conventional approaches. Here they are:

Definition: Agile practices

> *Agile practices are approaches to develop and implement solutions to achieve organisational goals in unplanned situations under uncertainty. They are characterised by independently working groups of competent individuals which use special agile procedures. These agile procedures are typically iterative. In principle, agile practices can be informal or predefined.*

Notes on the definition of "Agile practices"

Note 1: *Independently working groups of competent individuals are characterised by the fact that they act as a group in a self-directed and autonomous manner. This means that they are at least free to choose the methods of working on the solution and to make decisions independently without having to name leadership functions or to involve further responsible persons. These groups are also called self-organised teams, self-directed teams or **agile teams**.*

Note 2: Iterative procedures *are characterised by the fact that the specific approach is not planned into detail from the very beginning, but is developed gradually. Short decision and feedback cycles are the key factor, so that the path taken is regularly reviewed and unfavourable effects or unwanted results can be corrected at short notice.*

Note 3: *Examples of agile practices include self-directed ad hoc working groups, stand-up meetings, and service design labs.*

Note 4: Agile methods *are special cases of agile procedures, which are defined for predetermined use cases and can be applied in a standardised way in different kinds of organisations. Examples of agile methods are Scrum, Kanban Boards, and Extreme Programming.*

In this way, agile practices can be clearly distinguished from a chaotic, somehow flexible approach. Firstly, a VUCA problem is needed, which is to be solved in the sense of the organisational goals. Secondly, the right people are needed, because an agile team consists of (for the respective problem) competent individuals. Thirdly, iterative procedures are used for gradually reaching an optimal solution under uncertainty.

Note: Agile practices solve VUCA problems and are characterised by self-directed agile teams and the use of agile iterative procedures

These aspects are explained in more detail below, with regard to their practical application.

1.2 The Agile Team: Quality Circle Reloaded

Agility is always a team task. A single person can be flexible but not agile. A truly agile team is characterised by the fact that it can decide independently. This no longer requires a manager. Self-control extends to what the team does (goal setting), how it performs the task (methodology and tools), who does what (task assignment), and when it is finished (time constraints). This simply makes a manager who traditionally decides most of the above-mentioned aspects dispensable.

The basic idea is that in the VUCA world, even a very well-informed manager is not able to keep track of all relevant aspects for a good decision. Thus, she would have to consult with the experts in the team anyway. A team decision could only be optimal when all the necessary skills and perspectives are on board. In practice, however, these requirements are often not respected. Locking up any kinds of people who just have some time to spare in a design lab and prompting them to find an innovative solution will rarely produce a meaningful result. Even if managers promote agility with the statement "everybody be agile now" they still need to dedicate fitting resources—in quality and quantity. As important is the willingness of the agile team members to accept the responsibility and their ability to use iterative, agile approaches.

Example: Rescue service

There are industries where self-controlling teams are not new. When an emergency response vehicle arrives at the scene, there are usually two paramedics or in some cases a doctor and a paramedic on board. Treatment at the scene of the emergency is carried out by the rescue teams in mutual agreement and ranges from the administration of medication to the stabilisation of the patient's circulation or to intensive medical treatment such as artificial respiration. Emergency rescue is always a team task. To ensure the competence of the team members there is intensive special training, which is kept up to date with at least 40 h of professional development per year (Rettungsdienst-Kooperation in Schleswig-Holstein (RKiSH) gGmbH, 2019b). Emergency rescue operates in a typical VUCA environment: Location, patients, and situation are constantly changing and cannot be influenced by the rescue service—in other words, far away from the controlled conditions of classic standardised production processes. The RKiSH, for example, has been certified according to ISO 9001 since 2008 (Rettungsdienst-Kooperation in Schleswig-Holstein (RKiSH) gGmbH, 2019a).

For larger agile teams, such as those used in development processes, it is very helpful to use an experienced facilitator. The aim is to support the work process with the help of suitable tools. Such a facilitator ensures that appropriate agile tools—i.e. approaches and methods—are used and that team decisions and their adherence are made transparent to all members. For this, the facilitator needs a toolbox of agile approaches and personal experience with coordination processes in agile teams in order to ensure that the team works efficiently.

Quality managers are no strangers to dealing with agile teams. After all, the beginning of the quality management (QM) movement was the realisation that it makes sense to have improvement ideas realised by those who have direct experience with the related processes. Quality circles were originally developed for exactly this purpose. That is why workers were trained in the Q7 quality tools. Viewed in this light, agile teams represent only a consistent enhancement, a kind of "quality circle reloaded".

Yet for most organisations, the use of true agile teams without a leader is revolutionary. For more information on the practical challenges, see Sects. 3.2 and 3.3.

1.3   The Work Process of Agile Teams

The specific work process of an agile team can look very different. The starting point for agility is often marked by ad hoc working groups, which independently take on urgently needed improvement measures or work on special assignments. The team clarifies the topic for itself, chooses the frequency of meetings as well as the procedures used, and is free to choose one of the solutions developed. A truly agile team also receives the necessary resources to implement the solution created immediately. In order to form these teams, everyone who is familiar with the topic to be dealt with and who can and want to contribute important perspectives is brought on board. First, essential aspects and ideas are collected, usually with the help of general management, quality, or creativity management tools. Agile teams are characterised by the fact that they work less with tried and tested standard methods (e.g. Q7, M7) and more with unusual visual and co-creative methods. These include visual facilitation or Lego Serious Play. That is why more and more companies are setting up agile workspaces with lofty titles like "Design Lab", "Agile Workspace", or "Creativity Room". The spatial concepts behind it promote interaction and group work. They deploy flexible furniture such as rollable tables and chairs, movable walls, sofas, and beanbags for different settings. In this way, large-scale group work, work in small or breakout groups, and individual work can alternate on the fly. Also, it is possible to choose a preferred operation mode—be it while sitting, standing, or sometimes even while lying down. In addition, there are extensive recording options, such as write-on walls, whiteboards and electronic whiteboards, flipcharts, pin boards, and the even more flexible Kapa boards. In addition, the rooms already contain all the necessary materials, such as Lego bricks, modelling clay, six hats, post-its, or picture cards.[4] The more experienced an ad hoc working group is with using agile approaches, the more varied the selection of methods used and the work process as a whole becomes. This also shows once again that it can be very helpful to use experienced facilitators who are familiar with agile work processes and co-creative methods. They also ensure that the team

---

[4] Further information is available from Klaffke 2019.

determines when and how to make decisions. Frequent and regular coordination and selection cycles are typical for iterative, agile approaches. In order to enable these, non-prioritised ideas and solutions are not discarded, but are captured in the form of a "parking lot" or topic backlog instead. In this way, they could be used later, should a previously prioritised idea prove to be inappropriate or not feasible in subsequent stages. In such a rather mild form, agility is easily recognisably as "quality circle reloaded". It becomes considerably more challenging when core processes are reorganised in such a way that self-controlling teams take the place of standardised manufacturing processes that were originally organised based on a classical division of labour.

Example: Individual customer order processing

One of the companies surveyed manufactures around 10,000 different product varieties based on roughly 30 basic products through made-to-order and thus individual production. They promise their customers that their order will be processed within 24 h. To achieve this, the different departments previously involved in order management were dissolved. Cross-functional agile teams were formed instead. These teams decide for themselves which orders they take on and process. The employees help each other out—within a team and, if necessary, across teams. No one goes home until it is all "done". In order to enable the employees to support each other, they have been and are being intensively qualified. Their possible fields of applications are recorded in a qualification matrix.

Stand-up meetings also belong to the list of typical agile procedures. To ensure that all employees are informed about the work progress of their team colleagues, for example, every morning all employees meet for 15 min while standing. Everyone reports briefly and concisely about what they have achieved and where they may need some support. In this way, the team coordinates its activities and keeps an eye on the achievement of its objectives. The tight time frame helps to concentrate on the essentials. For a detailed reflection on what has been achieved or a fundamental reorientation, there are other complementary meeting formats available.

Example: Stand-up Meeting

The leader of an HR department decided to be a frontrunner of agility. Thus, he introduced daily stand-up meetings in his team. Every work day, his whole team met for 30 min in order to discuss the daily tasks. After 2 weeks people started to feel that it was useless. "We did not know what else to discuss", one of the team members stated. No wonder, as the team leader followed the classical leadership approach. Therefore, he distributed the work and coordinated the usual activities. A stand-up meeting is a useful coordination mechanism for a self-organised team that needs the daily information to synchronise its tasks. In case the leader serves as the coordination mechanism, such a meeting is pointless. It is also a waste of time, as it is faster to assign tasks on a one-on-one basis without the others looking on.

In a real VUCA environment a detailed planning and definition of activities in advance is not possible in a meaningful way, because not enough reliable information is (can be) available. Therefore, iterative, step-by-step detailed processes work much better here than extensive process instructions and detailed planning, which only provide an advanced planning security. Nevertheless, dealing with these iterative processes is a challenge, as shown in Fig. 1.3.

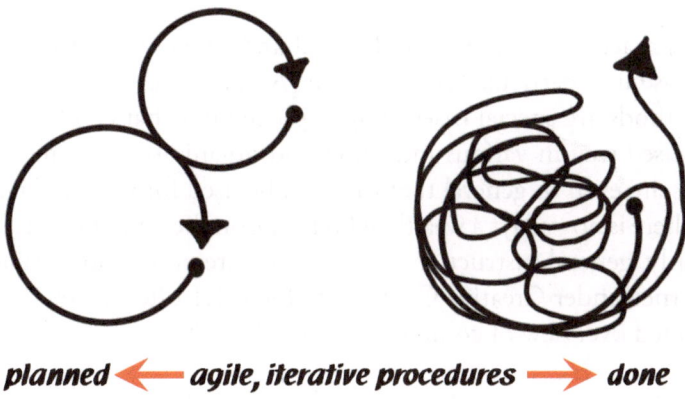

planned ⟵ *agile, iterative procedures* ⟶ *done*

Fig. 1.3 Special features of iterative procedures (own illustration)

Example: Individual assessment of oil quality

Another company surveyed deals with natural oils of various kinds. These natural products—in particular in case of organic cultivation—can vary greatly in quality from one year to the next. They keep close contact with suppliers all over the world. Regularly, their purchasers visit the farmers who grow the raw materials and discuss with them what quality the oil is expected to have. At the same time, they widen their global network of potential buyers. Their account managers discuss possible orders for oils and gather requests for special kinds of oils with extraordinary uses. Regularly, purchaser and suppliers meet to exchange their knowledge gathered worldwide and discuss their ideas. This is completely self-organised by the team and not coordinated by appointed managers.

As there is nowadays an ever-growing demand for oils as raw material for organic packaging and other sustainable solutions, they bridge the gap between the strategic ideas of the buyers and possibilities of the producers. This might even mean that they develop a supplier for years towards a certain demanded quality. Typically, after an oil is offered, it is chemically examined in detail in their very own high-tech laboratory. Afterwards, they create a mixed agile team in order to consider possible uses for the exact quality and quantity of oil they were offered. In case of very extraordinary oils, they include customers and producers in their teams. Their unique approach ensures their success on an ever-changing world market.

1.4 Agile Methods: Kanban Boards, Scrum, and Co

Agile approaches can be very individual and often evolved from a very specific use in a particular environment. As already defined in Sect. 1.1, agile methods are special cases of agile procedures that can be used in a standardised way in various industries and organisations. Some of these methods are even so general that they can be used for very different purposes. There is nowadays a wealth of literature on these methods available.[5] Frequently, general instructions for their use are even made available on the Internet under Creative Commons. Table 1.1 offers a very individually selected overview of common agile methods.[6]

---

[5] Amongst other things, the study of Komus and Kuberg 2020 deals with the importance of individual agile methods, admittedly with a strong focus on software development.

[6] For a more in-depth study of agile methods, further sources are recommended. For agile learning formats especially Graf et al. (2022). For Lego Serious Play, detailed information on the methodology is available from Blair and Rillo (2016). For Design Thinking the following is highly recommended: Lewrick et al. (2018).

Table 1.1 Agile methods and their area of application—selected examples

Agile method	Field of application	Definition	Further information
BarCamp	Continuing Professional Education (CPE)	Agile conference format ("un-conference" for large groups). Only the main topic is pre-defined. Individual topics for discussions and workshops are suggested, selected and facilitated by the participants themselves. (This format is very similar to Open Space conferences)	http://www.barcamp.org/
Brown Bag meeting	CPE	Self-organised learning format. Interested employees meet for about 1.5 h in a seminar room for learning about an announced topic. One employee introduces the topic to the others and answers questions. The participants can eat the food they have brought with them (in their "brown bags", hence the term)	https://www.investopedia.com/terms/b/brown-bag-meeting.asp
Design Thinking	Development	Innovation method. Structured, iterative approach to generate creative ideas using a variety of design methods. The starting point is always the user perspective	https://hpi.de/en/school-of-design-thinking/design-thinking/background/design-thinking-process.html
Extreme Programming	IT/software development	Agile process model for software development emphasising simplicity and customer orientation. Integrated practices such as pair programming are carried out in timed intervals. The code is created section by section and optimised through continuous testing ("test-first approach")	http://www.extremeprogramming.org
Kanban Board	Team organisation	Tool for visualizing the workflow of a team, e.g. in projects. It creates transparency about the progress of work and any bottlenecks that may arise. It facilitates targeted resource management via work-in-progress limits (see Fig. 1.4)	https://www.scrum.org/resources/kanban-guide-scrum-teams

(continued)

Table 1.1 (continued)

Agile method	Field of application	Definition	Further information
Kickbox	Innovation/CPE	Agile innovation format for individual application or as a learning format in the form of team kickboxes. The kickbox contains notes and work packages that pre-structure the learning or innovation processes of the users. Originally developed by Adobe to promote innovation	https://kickbox.org/
Lego Serious Play	Development/ teambuilding	Method for generating ideas. In a structured, facilitated process, ideas or solutions are designed as Lego models. The team-based approach is based on storytelling: everyone builds, everyone tells the related story. The focus is always on the Lego models ("identities") that were built	https://davidgauntlett.com/wp-content/uploads/2013/04/LEGO_SERIOUS_PLAY_OpenSource_14mb.pdf
Scrum	Development	Incremental process model for development projects, especially in software development. The development is carried out module by module in time-limited sections (so-called sprints). This approach uses three clearly defined roles (product owner, Scrum master and Scrum team) on the basis of a detailed, customer-centred and transparent process	https://scrumguides.org/
Working Out Loud (WOL)	CPE or individual development	Self-directed peer support approach. Four to five people come together and form a Working Out Loud Circle. Members support each other in achieving their individual development goals. The circle meets every week for one hour that is facilitated according to a pre-structured process based on so-called Circle Guides. All members work on their individual goals. The meetings will be held for 12 weeks in physical presence or virtually	https://workingoutloud.com

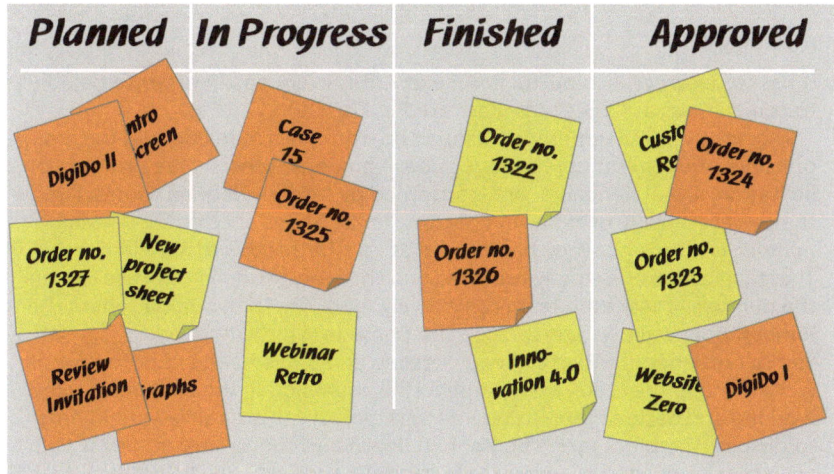

Fig. 1.4   Kanban Board (own illustration)

Experienced quality managers or facilitators are familiar with some of these methods for years. Some methods were derived from older concepts of organisational design such as lean management or from the pool of creativity techniques. This underlines the view that agile work is not really new. On the positive side, the new willingness of organisations to consequently embrace agility (i.e. introduce self-organised agile teams and iterative agile procedures) offers some of these methods a framework in which they can develop their full potential for the first time (Fig. 1.4).

Example: Kanban Board: How (not) to do it

In some organisations, agile methods are adopted without really understanding their basic ideas. A typical example is the widespread use of Kanban boards. The simple basic principle that everyone searches for the next work package ("pull") and adds a card to the status overview for tasks currently being worked is often used in agile (project) teams. So far, a Kanban Board is only a visually spiced-up "team to-do list". Quite useful, however lastly only a simple whiteboard with some information on it, which can be realised as a real board on site or electronically (e.g. by use of Trello or Jira). It is surely nice to be informed about the current work status of the team by means of such a board. However, it does not help a team if

(continued)

(continued)

it has to decipher a colourful jumble of notes every day in order to be able to come across as "agile" (as illustrated in Fig. 1.5).

The rightly popular method "Kanban Board" offers more. Besides the basic principle of visualisation, it also includes—amongst others—the principle of limitation. A real Kanban Board is governed by a clearly defined control logic, the so-called WIPs or work-in-progress limits. These specify the maximum number of tasks or work packages that can be processed simultaneously. These limits must be strictly observed and thus constitute an effective cap on the number of tasks the team can actively work on. The visualisation on the Kanban Board allows users to see the phase in which the work is jammed. Such a bottleneck blocks further processing and prevents the smooth flow of work through the team. In response to the visual signal, the team has to stop working and decide upon fitting interventions. Those are carried out in order to dissolve the blockages. The Kanban Board also shows very quickly if such an intervention has not been fully thought through and is ineffective. In these cases, the jam is only shifted to subsequent steps. In practice, this happens, for example, when non-expert employees are employed in order to speed up the process. They often make errors that only become apparent during the final check and then have to be corrected laboriously.[7] Used correctly, a Kanban Board unfolds a very convincing added value.

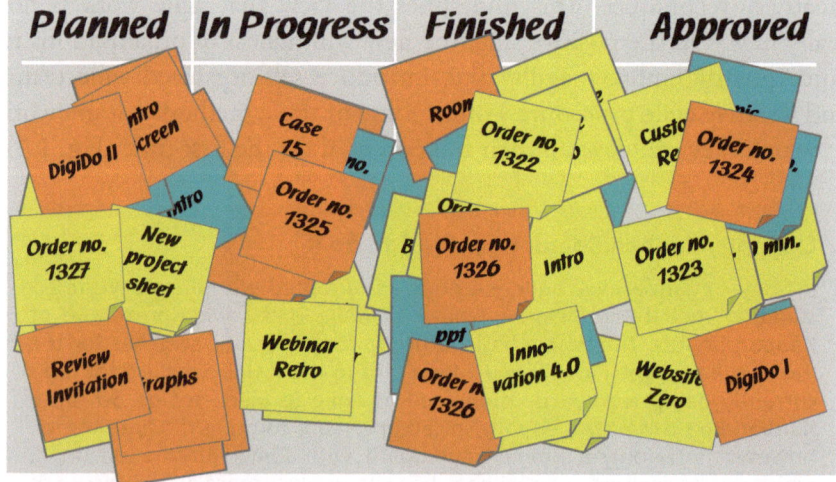

Fig. 1.5   Kanban Board—How not to do it (own illustration)

---

[7] Hanschke (2017), pp. 22–24, Eisenberg (2018), pp. 163–164. The latter book shows the possibilities of using Kanban in a very entertaining way.

1.5 Processes, Projects, and Agility: The Agony of Choice

Organisations have various possibilities to shape the fulfilment of their tasks and thus the achievement of their objectives. A design based on processes or projects is common. A process organisation (in analogy to ISO 9000:2015, 3.4.1) defines a company on the basis of activities it usually carries out, i.e. as a set of interrelated or interacting activities. A project organisation is useful for temporary tasks, undertaken to achieve objectives conforming to specific requirements (ISO 9000, 3.4.2). These tasks usually have a unique, innovative character and considerable complexity. Now, the third variety is an agile organisation. It uses agile practices to optimise task fulfilment under special environmental conditions, namely the VUCA world. Although agility enthusiasts like to propagate the idea that agility is the one and only form of designing tasks for modern organisations, all three forms of organisational design can still be justified. In the real business environment, all forms of design coexist, depending on their specific purpose.

In addition, the determining elements of these three forms of design can be combined—not only in the overall view of the organisation but also in the design of individual activities (as depicted in Figs. 1.6 and 1.7). A project can, for example, be designed in the classic way as a "waterfall" and thus highly process-oriented. But it can also be designed as an agile SCRUM project. In most organisations, mixed forms will be found, which contain determining proportions of all three varieties. The classification of the organisational design resulting from these three options can be illustrated using a "system cube", as presented in Fig. 1.7.

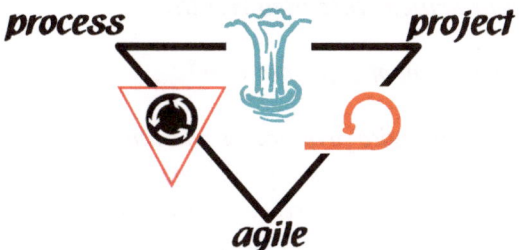

Fig. 1.6 Three principles of organisational design (own illustration)

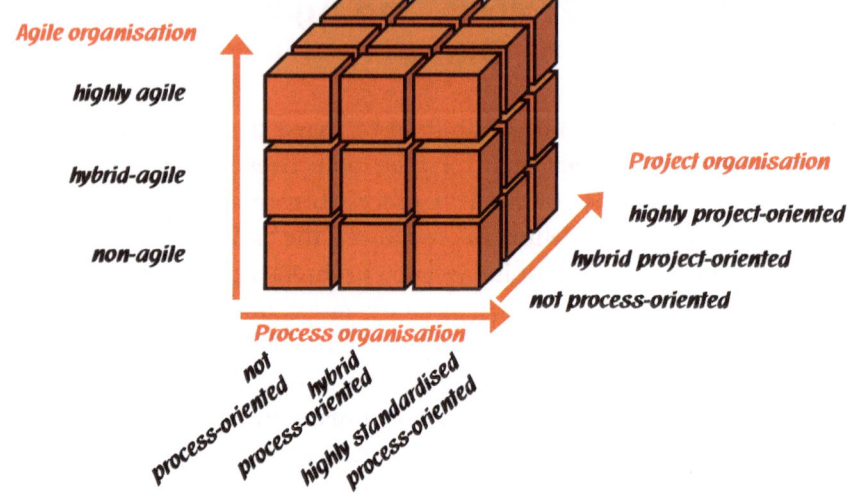

Fig. 1.7 System Cube—three principles combined (own illustration)

Just as a single project can combine project and process orientation, organisational processes can be combined with agility, thus creating agile processes. In these cases, the activities are represented and designed as a set of interrelated or interacting activities, but certain of these activities or (sub-) processes are carried out using agile practices. Seen from a quality management point of view, this becomes significant, if it deviates from the basic idea of a controlled process. This consideration leads to the following definition of agile processes.

Definition: Agile processes

Agile processes are processes that use agile practices to a relevant extent to determine and achieve intended results.

Notes on the definition of "agile processes"

Note 1: *The **extent of agile practices** is relevant if it changes the nature of the process in such a way that dealing with uncertainty and constant change in a self-directed manner takes precedence over fixing controlled conditions. This does not necessarily mean that activities using agile practices actually predominate in terms of time, resources, or value added.*

*Note 2: The **determination of intended results** during the process does not mean that there are no requirements from interested parties at the start of the process. However, at the beginning these are not defined in such a manner that the process can be aligned with them. This is often the case in innovation and development processes. Processes with a high extent of agile practices for determining anticipated results are typically iterative.*

Note: Agile processes are naturally dealing with uncertainty and constant change instead of relying on controlled conditions. They use agile practices to a relevant extent.

References

Adam, P. A. (2018). *System(at)isch agil. Wie agile Prozesse in ein Managementsystem nach ISO 9001:2015 integriert werden können*. Whitepaper. Hochschule Hannover. Fak. IV, Abt. BWL, Management series no. 1. Accessed October 14, 2022, from https://doi.org/10.25968/opus-1268

Blair, S., & Rillo, M. (2016). *Serious Work. How to facilitate meetings & workshops using the Lego® Serious Play® Method with conscious incompetence*. ProMeet.

Eisenberg, F. (2018). *Kanban – mehr als Zettel*. Carl Hanser Verlag GmbH & Co. KG.

Graf, N., Gramß, D., & Edelkraut, F. (2022). *Agiles Lernen. Neue Rollen, Kompetenzen und Methoden im Unternehmenskontext. 3. Auflage 2022*. Haufe Lexware (Haufe Fachbuch).

Hanschke, I. (2017). *Agile in der Unternehmenspraxis*. Springer Fachmedien Wiesbaden.

Information is Beautiful. (2015). *Codebases. Millions of lines of code, v 0.9*. Accessed October 14, 2022, from https://informationisbeautiful.net/visualizations/million-lines-of-code/, updated on 9/24/2015.

Klaffke, M. (2019). *Gestaltung agiler Arbeitswelten*. Springer Fachmedien Wiesbaden.

Komus, A., & Kuberg, M. (2020). *Result Report: Status Quo (Scaled) Agile 2019/20*. 4th International Survey – Benefits and challenges of (scaled) agile approaches. With assistance of BitKom, GPM, IPMA, PMA, Scrum. org, SPM, Swiss ICT. Hochschule Koblenz, BPM-Lab for Business Process Management and Organizational Excellence. Accessed September 22, 2022, from https://www.process-and-project.net/studien/download/downloadbereich-status-quo-scaled-agile-2019-2020/

Lewrick, M., Patrick, L., & Leifer, L. (2018). *The design thinking playbook. Mindful digital transformation of teams, products, services, businesses and ecosystems.* With assistance of Nadia Langensand. Wiley.

Rettungsdienst-Kooperation in Schleswig-Holstein (RKiSH) gGmbH. (2019a). *Ein Rettungsdienst mit Charakter.* Accessed September 22, 2022, from https://www.rkish.de/unternehmen/wer-wir-sind/die-rkish.html

Rettungsdienst-Kooperation in Schleswig-Holstein (RKiSH) gGmbH. (2019b). *Für den Notfall gerüstet.* Accessed September 22, 2022, from https://www.rkish.de/notruf-112/wer-kommt/das-rettungsteam.html

Scheller, T. (2017). *Auf dem Weg zur agilen Organisation. Wie Sie Ihr Unternehmen dynamischer, flexibler und leistungsfähiger gestalten. 1. Auflage.* Verlag Franz Vahlen.

Sommerhoff, B. (2016). *Manifest für Agiles Qualitätsmanagement. DGQ Deutsche Gesellschaft für Qualität.* Accessed October 14, 2022, from http://blog.dgq.de/manifest-fuer-agiles-qualitaetsmanagement/

The Agile Alliance. (Ed.) (2001). *Manifesto for Agile Software Development.* Accessed October 14, 2022, from http://agilemanifesto.org/iso/en/manifesto.html

2

Agile in ISO 9001: How Does That Work?

Abstract In this chapter, specific examples are used to explain how agile practices can be integrated into a management system according to ISO 9001:2015. Particular emphasis is placed on the assumption of planning and monitoring activities by self-controlling agile teams, the control of agile processes with the help of the so-called BIG FIVE, and the creative documentation of agile practices.

Keywords Agility • Agile practices • Agile process • Agile teams • ISO 9001 • Quality management • Process control mechanisms • Documentation

2.1 The Grey Area Between Standard Process and Chaos

If we want to find out how much agility a certified QM system according to ISO 9001:2015 can tolerate, the first question that arises is whether ISO 9001:2015 fundamentally limits the choices concerning the organisational design. Is there a clear guidance on the use of process, project, or

© The Author(s), under exclusive license to Springer Nature Switzerland AG 2023
P. A. Adam, *Agile in ISO 9001*, Business Guides on the Go,
https://doi.org/10.1007/978-3-031-23588-7_2

agile organisational design? A glance at the standard quickly shows that there are no specifications whatsoever with regard to projects. The use of projects is neither explicitly required nor prohibited, although ISO 9000:2015 offers a definition of projects (3.4.2). Accordingly, it is not surprising that the design of temporary activities with a substantially unique character is usually carried out as a project, even in organisations featuring certified QM systems. However, with regard to the remaining varieties—process-oriented and agile organisational design—some considerations and requirements can be obtained from ISO 9001:2015. Figure 2.1 represents the following contemplations graphically.

The so-called process approach is one of the principles of quality management according to ISO 9000:2015 (section 2.3.4). It is assumed that consistent results are achieved more efficiently and effectively if the activities are understood as interrelated processes and these in turn are managed as a coherent system. This approach is taken up by ISO 9001:2015 and is particularly developed with specific requirements in section 4.4 (DIN EN ISO 9001:2015-11). Amongst other things, the organisation must determine the inputs, outputs, sequence, and interaction of its processes. In addition, it is necessary, inter alia, to identify process inputs and outcomes,

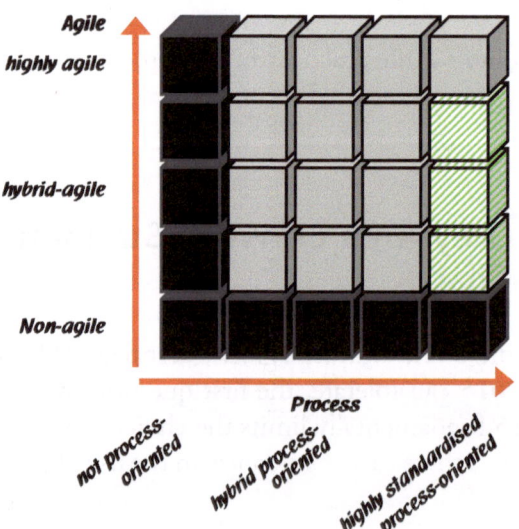

Fig. 2.1 Grey areas of organisational design according to ISO 9001:2015 (own illustration)

and to define and apply the criteria and procedures to ensure the effective implementation and control of these processes. With regard to the core process, section 8.5 states that production and service provision shall be carried out under controlled conditions. In addition, section 8 determines various details, particularly with regard to the control of certain types of processes. From this it can be concluded that an organisation that is not (at all) process-oriented is not compatible with ISO 9001:2015. This makes a lot of sense from an organisational design perspective. For example, when it comes to standardised accounting processes such as annual financial statements, every association and corporation is required to execute them in a comprehensible and controlled manner. This already results from requirements of shareholders and investors as well as tax and legal regulations. If the context is well known and the respective situation can be clearly specified, solutions for recurring problems exist. These can be described in a standardised way and easily optimised in terms of effectiveness and efficiency. In these situations, a process-oriented organisation is always the most effective and efficient way to handle these types of tasks. For example, the production of CDs is a contamination-sensitive activity. Therefore, it takes place under precisely defined humidity, temperature, and air pressure conditions in cleanrooms. The contamination of air is controlled through an air treatment system and of surfaces through abrasion-resistant stainless-steel constructions. Employees have to comply with very strict conduct requirements. They have to wear protective clothing and go through intensive cleaning routines before and after entering the cleanroom. The employees need to be trained to use slow and study movements, as making rapid or frantic movements carries additional risks. In such an environment, an agile re-organisation would neither be necessary nor suitable. Highly standardised process-oriented organisations correspond to (highly) controlled conditions. They certainly comply with the requirements of ISO 9001:2015 and are therefore "in the green zone". The grey zone arises in the hybrid areas of organisational design: It has to be decided how much process orientation or which minimum of "controlled conditions" is sufficient to design a reliable organisation and enable an ISO 9001 certification.

ISO 9001:2015 does not mention agility or agile practices. However, the standard contains concepts that reflect the essential values and principles of the Agile Manifesto (The Agile Alliance, 2001). According to

ISO 9000:2015, customer focus, engagement of people, and (continuous) improvement are amongst the essential principles of any effective QM system (DIN EN ISO 9000:2015-11, pp. 13–18). Consequently, ISO 9001:2015 contains various requirements that shape these principles. First of all, this includes customer focus (section 5.1.2). This is a principle that is also reflected in the Agile Manifesto with top priority. In addition, there are employee-related aspects of ensuring competence (section 7.2.b), awareness of one's own contribution (section 7.3.c), and defining relevant internal and external communications (section 7.4). All these principles show a high overlap with the agile value "individuals and interactions". Of particular relevance in ISO 9001:2015 is improvement (section 10). It is very close to the agile value "responding to change" and the principle of reflection in regular intervals. The determination of risks and opportunities is the core task of quality management in the form of "risk-based thinking". It must be applied when selecting suitable improvement measures and determining opportunities in accordance with ISO 9001:2015, e.g. in order to be able to meet future customer requirements and expectations or to improve performance. If we assume that the postulate of the VUCA world is justified, it follows that an organisation which completely dispenses with any kind of agility would not be certifiable. However, there is uncertainty about how much agility and what kinds of agile practices an organisation can tolerate before agility comes into conflict with other ISO 9001 requirements and thus jeopardises certification. It is important to find out where the green area ends and the risky grey area begins.

In the overall view, "guard rails" must be found for the green-hatched and light-grey areas shown in Fig. 2.1, so that they could be clearly distinguished from the black—and thus non-certifiable—areas. The QZ article by Adam (2021) already contained five practical tips for an ISO 9001-compliant design of agile practices (Adam, 2021, p. 47). In the following chapters, these are developed in more detail.

Note: Every organisation can freely combine the design principles process, project, and agile. In certified QM systems a minimum of process orientation and agility is indispensable. Every practical approach has to explore the acceptable grey areas.

2.2   Planning, Controlling, and Monitoring

The requirements of ISO 9001 are manifold. For dealing with agility, it is particularly relevant that an organisation must plan, control, and monitor the entire QM system with its risks and opportunities and necessary resources, as well as the procedures in the core processes ("operation"). In order to determine the conditions under which agile practices are compatible with ISO 9001:2015, two questions have to be answered first:

- Are agile self-organised teams actually allowed to take over planning, controlling, and monitoring activities?
- Are planning, controlling, and monitoring activities in line with agile procedures?

ISO 9001 and Agile Teams
In most organisations, planning, steering, and monitoring activities are typical management tasks and are thus carried out either by managers or by specific departments. These special departments typically include quality management or process management. However, these practices are not based on explicit requirements of ISO management standards. ISO 9001:2015 does not specify who is allowed to carry out these tasks. In principle, any function or hierarchical level could carry out these activities, provided that the responsible employees have the necessary skills. The standard always requires that any kind of internal regulation must be adequate and functional. In addition to the competencies of the parties involved, the type of activity under consideration, the process objective, and the associated risk profile are crucial factors for evaluating the appropriateness of an approach.

Thus, it is not to be rejected in principle if an agile team, as part of its self-organised approach, independently sets its goals, determines the type of procedures to be employed, divides the tasks amongst the team members, and, after completion, assesses the achievement of the goals and, if necessary, develops, prioritises, and initiates improvement measures. Since the revision of ISO 9001:2015, the principle "output matters" has been applied in many areas. This means, as long as it can be proven that the objectives are reliably achieved over a longer period of time and the results meet the requirements of the customers and other interested parties involved, even unusual internal regulations can be certified. Ultimately, the motto is: "If it works it works."

Note: Agile teams are no problem for ISO 9001!

Example: Development with the Scrum method

The heart of the Scrum methodology[1] is the so-called Sprint: A fixed period of time of maximum 4 weeks in which a precisely defined product goal (which is usually not a complete product but a module of the overall product to be developed) is completed. In such a Sprint, the necessary planning, controlling, and monitoring activities are defined from the outset in the form of further so-called Scrum events. The aim is to enable inspection especially at critical points and to create transparency as an actual implementation of the "inspect and adapt" approach. One of the Scrum events that the whole development team (=Scrum team) performs together right at the beginning is the Sprint Planning. This event initiates the Sprint by defining the Sprint Goal to be delivered and the work required for this, broken down into small units. In daily 15-min short meetings (Daily Scrums) of the development team, progress is reviewed and adjustments and rescheduling are made if necessary. At the end of the Sprint, a Sprint Review takes place, to which the Scrum team usually invites other stakeholders. Here, the achieved status of the Sprint is reflected and transformed into input for the upcoming Sprint Planning. Budget, planned product features, and the schedule for the product release are reviewed and, if necessary, completely revised. In an additional Sprint Retrospective before the next Sprint Planning, the past Sprint is reflected in terms of the processes, the tools used, and the collaboration of the people involved. The goal is to create a plan for improving the (internal) functioning of the Scrum team.

The experiences of several interview partners with these Scrum events were universally positive. The very open critical Sprint Reviews were especially emphasised. Due to the close cooperation in the team, there is a greater willingness to address problems clearly and solve them consistently. Review processes in classic development projects tend to be concerned with exposing colleagues to other departments or the managers responsible for the reviews. In self-organised Scrum teams, however, there is a very sincere working atmosphere. The result is that all team members honestly identify cooperation problems and focus on improving the team's internal work processes without assaulting team members on a personal level. The result-focused guidelines of the Scrum methodology heavily contribute to this. Overall, it can be stated that review processes of agile teams are often even superior to classical review processes.

---

[1] All details of the methodology are presented in the so-called Scrum Guide; see Schwaber and Sutherland (2020).

Planning, Controlling, and Monitoring of Agile Procedures

It has already been clarified that agile practices are far from a "we do something with somebody somehow" approach. In the interviews it became clear that agility is about solving VUCA problems and not about a "Port of Freedom". As one of the interview partners put it: "Agile teams are not artists, after all". Ultimately, the aim is to achieve a (business) goal. This is a prerequisite for the use of agile practices in any organisational context.

Any kind of agile cooperation in a self-organised team will only be successful if it is coordinated well. Thus, coordination is essential for all agile teams. The necessary coordination is given, if all members of the team know exactly what they are responsible for. Clear roles and responsibilities have to be defined, as well as the expected results of the activities pursued. In software development, for example, this could be organised as follows: One colleague defines the interfaces to other relevant systems, a second programs the access rights, and a third performs all test activities. A Kanban Board might hang behind them and provides full transparency concerning who is working on what at any given moment. Problems are discussed and solved together in a daily stand-up meeting, usually first thing in the morning. If one task is held up, the members of the agile team help each other out—as long as they have the necessary skills for the task at hand. In these cases, they decide as a team whether they will look for a new task after completing their previous one or whether it is more important for the joint achievement of objectives to first support other colleagues that are stuck. Such a decision can only be made in a well-founded manner if the goal and the state of affairs are transparent for everyone.

During the execution of planning, controlling, and monitoring activities special emphasis has to be laid on the required level of detail. Due to their iterative, step-by-step approach and the very short (usually daily) review cycles, an agile team does not need an extremely detailed plan at the beginning. Instead, the degree of detail increases with the progress of goal attainment. Therefore, with regard to the work process many tasks are not detailed out before these jobs are actually taken over. It is to be noted that ISO 9001 does not contain any general specifications concerning the level of detail. Here again, the principle "if it works it works" applies. For VUCA problems, detailed planning at the beginning is wasteful anyway (see Sect. 1.3).

Note: ISO 9001 requires that planning, controlling, and monitoring activities are carried out reliably. In a VUCA environment, this can be ensured through agile practices.

2.3   Process Control: The BIG FIVE

Process control means that certain activities are carried out to ensure that the intended process result is produced. The definition of process control from the research project is as follows.

Definition: Process Control

Process control is a system designed to ensure that the process result corresponds to the intended result. This system employs special mechanisms that are consistently applied in a control loop.

According to the requirements of ISO 9001:2015, section 8.1, operational planning and control activities include in particular the determination of requirements (a), the definition of criteria for processes (b), the determination of necessary resources (c), the implementation of control of processes in accordance with the criteria (d), the documented information for the corresponding evidence (e), the monitoring of planned changes, and the assessment of consequences of unintended changes. For certification, it must be demonstrated that the control system really works based on the performance of these and other activities.

Over the last few decades, typical mechanisms for process control have been established in all kinds of management systems, which are generally considered appropriate by auditors. These include statistical process control via Shewhart control charts in production environments or approval of payments by authorised persons in accordance with the four-eyes principle in finance. It is undisputed that especially statistical control measures are only useful for standardised processes in controlled environments and therefore cannot be applied in VUCA ecosystems. In some publications, the use of agile practices—where agile teams independently select and implement their agile procedures and help each other out if the need arises—is understood as an independent concept and counter-model to the usual process-related control. Under these premises, the critical questions are:

- Can agile processes really be controlled in a QM-compliant manner without losing their character?
- If yes, to what extent?

Within the framework of this research project, a master's thesis was awarded for answering these questions. Tonio Japing has taken up this task. He investigated in his master's thesis whether abstract, uniform control mechanisms are perhaps hidden behind the multitude of process controls used in practice. With the help of such general process control mechanisms for ensuring the quality of process results, it would be easier for quality managers and auditors to classify even unusual control procedures and find out if they would be generally permissible. To this end, common control models from a wide range of processes and industries were examined, e.g. Statistical Process Control (SPC) and FMEA from the production area, the mechanisms of the "Internal Control System (ICS)" from the financial area, and the "Three Lines of Defence" from risk management. The large number of specific examples covered the complete range of functions of a typical company. All were examined for their commonalities and then consolidated on an abstract level. The resulting initially eight different high-level process control mechanisms were subsequently transferred to agile processes. It was possible to establish that the mechanisms investigated with regard to their function showed only marginal differences concerning the control of classical and agile processes.[2] In a nutshell, agile processes can certainly be controlled without agility suffering, and many mechanisms for designing agile work processes can very well be understood as ISO 9001-compliant process control.

---

[2] Japing (2018). The complete Master thesis (in German) is available online under Creative Commons.

Following up on this thesis, the findings were further deepened and consolidated. Finally, five fundamentally different process control mechanisms emerged that can be clearly distinguished. All of them can be applied to classic as well as agile processes, without exception. These are the so-called **BIG FIVE** of process control (see also Fig. 2.2):

1. Requirements

… undisputedly determine every process. They define what result has to be achieved, what goals are pursued with the individual activities, or what conditions the customer of the process requires for good execution. All this has a considerable influence on the way the process is designed and set up. According to ISO 9000:2015, 3.6.4, requirements can be generated by interested parties or the organisation itself and may be specified, generally implied, or mandatory. These requirements do not necessarily have to be in writing, although contracts, specifications, or other kinds of documented information constitute a typical format. They could also be obtained from a phone call or meeting with the customer. It is to be noted that requirements do not have to be available in full at the beginning of the process. They can be further developed step by step based on feedback obtained for drafts, prototypes, or samples. Such an iterative approach of obtaining requirements would be typical for agile development processes.

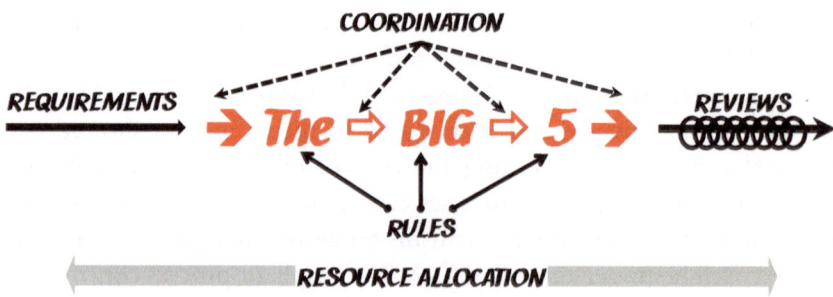

Fig. 2.2 The BIG FIVE of process control (own illustration)

2. Resource allocation

… has a recognisable direct influence on the quality of the processes and the achieved process results. Whether agile or classic, all necessary resources must be available. This applies to human resources taking two aspects into account: the appropriate quality, e.g. people with fitting competences and an appropriate working attitude, and the appropriate quantity, e.g. in the form of sufficient usable working hours. Resource allocation also includes the provision of financial resources or budgets, infrastructure, working environment, and management support. All necessary resources must be determined and provided according to ISO 9001, 7.1. Top management, in particular, is obliged to ensure the adequacy of the resource allocation (section 5.1). The optimal mix can vary greatly depending on the requirements. For example, accounting processes can be significantly supported by IT workflows and can therefore often be carried out faster and more error-free than without such a technical investment. Agile processes, on the other hand, are always highly dependent on human resources.

A special feature is the handling of the resource "**time**". In classic process environments, process control via time constraints is often geared towards meeting the dominant customer requirement of timely or rapid processing. Accordingly, an attempt is made to achieve time optimisation by controlling the individual activities, so that the process flow as a whole is accelerated or made more consistent. However, if time budgets are too tight, the defined process goals are easily missed, and sometimes employees are overburdened. In agile processes, especially in agile processes with highly creative work components, agile procedures are carried out in particularly tight time periods. This so-called time-boxing should enable concentration and focus on the task at hand. The primary goal is not to speed up the process as a whole, but to improve the result by increasing the focus of all team members involved. The origin of time-boxing was the realisation that allotting more time does not always lead to better results. The principle of time-boxing nowadays belongs to the standard repertoire of creative thinkers. It is used, for example, in Sprint events or in Design Thinking phases.

Example: Time-boxing in Sprint

All Sprint events are time-limited via time boxes. The Daily Scrum, for example, is set to a maximum of 15 min each day, so that the Sprint Goal is always in focus. The development team itself selects a suitable structure for the meeting time. The aim is to eliminate the need for other meetings, identify and remove obstacles, promote quick decisions, and exchange all essential information. It is quite common for individual members or even the whole team to schedule additional meetings afterwards in order to make necessary adjustments, reschedule other events, or discuss specific topics in detail (Schwaber & Sutherland, 2020, p. 9).

In "normal" meetings team members often encounter annoyances which massively reduce the efficiency of the meetings and frustrate the participants. For example, some people are late regularly and others have to wait. Some participants check their e-mails or talk to their neighbours in between, because they have the impression that the current topic is none of their business and they cannot contribute anything to it. In Daily Scrum, the strict time box ensures that such "idle times" are eliminated.

3. Rules

... determine which activities must be carried out within the processes and how the individual activities are to be carried out. Rules are often documented and become manifest in, for example, guidelines, codes of practice, instructions, policies, procedural requirements, or SOPs (standard operating procedures). Rules can have very different levels of obligation. For some processes, the oral definition that interns should always contact their internship supervisor if they have any queries could be sufficient. For others, extensive instructions are required. For example, the operation of SAP systems requires detailed manuals that explain in what cases which indicators have to be set. These have to be published and regularly updated. Of course, the same SAP manuals apply when members of agile teams use the SAP systems. Ultimately, an agile software programmer uses the same programming language as his classically working colleague. Although the paramedics of the rescue service make their initial diagnosis directly on site in an agile manner, the treatment itself is regulated (e.g. in Schleswig-Holstein) with checklists or standardised work instructions, which are available in a special small format for the trouser pocket. In most cases, detailed instructions for agile practices are inappropriate, as their purpose is to generate innovative solutions for unusual VUCA situations in a self-directed manner. Nevertheless, it may well be useful to establish general rules for these approaches as well. For example, you can specify that unusual assignments directed to the management accounting department are always the responsibility of the person who received the request. You can further specify that in these cases, the agile team must be composed of employees from all financial departments concerned, whereby at least one team member must have experience in facilitating agile teams. However, the agile team has to be free to individually define and coordinate the agile procedures used during the work process to correspond to the request.

How naturally agile procedures adhere to rules can be seen especially in the frequently used agile methods. These are popular precisely because they contain detailed rules of application. The Scrum Guide for example has 14 pages and carries the subtitle "the rules of the game" (Schwaber & Sutherland, 2020). As one interviewed developer put it: "By using Scrum we are much more regulated and transparent than with the regular development process employed before".

Example: Developing an Agile Team

A German unincorporated association introduced a "service design lab" and started working with agile teams. In the beginning, they only had one person that had some knowledge of agile tools and techniques and was able to facilitate these teams. As this facilitator could not be available every time an agile team was formed, they decided to produce a set of rules based on the best practices so far established. In this guideline, a number of acceptable agile methods and tools for certain stages of the design process were set. One year later they realised that the familiarity with agile practices and certain tools had increased tremendously across the board. More people were willing and able to facilitate agile teams, and even came up with a broad variety of different tools and techniques they wanted to try out. Following an intensive discussion, they allowed a first team to deviate from the guidelines and use new methods for the intended design process. After completion, an intensive review process was performed that showed very satisfying results. Thus, they broadened the guideline and allowed the use of more tools and techniques. Also, more people were allowed to attend training for certain methods. Nowadays, many people are trained in a broad variety of agile procedures, so that every agile team that is built can draw not only from different subject-specific experiences but also from a broad know-how concerning manifold agile methods. Guidelines for agile design processes were abandoned; instead, a collection of good practices is available, which is regularly updated.

Note: Agile methods are especially rule-based and thus fit very well into an ISO 9001-based management system. This is exactly why they are so popular.

4. Coordination

… ensures that the activities to be carried out in a process are coordinated in such a way that the entire process flow is optimised. It is therefore a matter of coordinating, controlling, and readjusting the interfaces between the different (sub-) processes and activities. Such coordination functions are typically performed by specific organisational roles or ensured by technical support (e.g. IT-based workflow systems). Coordinative roles can be designed in very different ways. The most common form of coordination is the distribution of tasks by the respective manager. In projects, the work packages are coordinated accordingly by the (sub-) project management. The overall coordination of several sub-projects is carried out by the overarching project management and the steering committee in its regular meetings. In agile development projects the role of the Scrum Master, for example, is a coordinating role, ensuring that collaboration is optimised and that goals and product domains are understood by everyone in the Scrum team. However, the decisions on the specific division of work are made by the development team within the framework of its self-organisation.

The design of the coordination function is one of the essential differences between agile and classically working teams. Schedulers, expeditors, or planners are used in various industries to ensure smooth processes and optimum resource allocation, e.g. for route planning in forwarding agencies and courier services, for scheduling temporary employees in HR services, and for optimising machine running times in production. As already explained in Sect. 1.3, agile teams excel especially in the coordination of very complex processes (VUCA problems), as all necessary information and competencies for a meaningful coordination can be directly integrated. In this sense, many agile practices can be seen as a special form of scheduling, which allows (unusually) high degrees of freedom at team level.

5. Reviews

… take place after the process or the individual sub-processes have been run through and are used for a structured assessment of the actual process flow and the process results achieved. They usually take place as a review meeting and may also include a formal acceptance of the result. As an essential input for continuous process improvement, they enable process errors, defects, or inconsistencies to be recorded in an orderly manner and to initiate improvements for the next run.

Reviews are of particular importance in agile practices—and especially when using agile methods. As already explained in Sect. 2.2, Scrum requires the implementation of Sprint Review and Sprint Retrospective. In Design Thinking, reviews of individual process steps in the form of evaluation, improvement, and reflection phases form an integral part of the methodological workflow (Blatt & Sauvonnet, 2017, p. 93). Especially in creative development processes it is nowadays state of the art to distinguish very strictly between two types of states of mind: divergence and convergence. Diverging phases should allow for an as-broad-as-possible creative development of ideas. They require from every participant an extremely open and accepting mindset, so that all contributions are valued and used as a starting point for further exploration and idea generation. In contrast, converging phases require a critical and objective evaluation. Participants are now allowed to focus on predefined selection criteria and their fulfilment, in order to assess specific functionalities and finally select potentially best solutions. A constant alternation between the two phases within the framework of Design Thinking enables a step-by-step, continuous optimisation of the approach to solving the problem by the respective development team.[3] The entire development process based on Design Thinking envelops a macro-cycle from the first ideas, critical functionalities, and funky prototypes to the final prototype. In each of the separated phases, the team iteratively runs through a micro-cycle from customer-centric observation to idea generation and finally testing of the respective prototypes (Lewrick et al., 2018, pp. 30–37). In the tests, prototypes are analysed in a variety of ways depending on the chosen tools and techniques. Expressed in ISO jargon: Prototypes are subjected to various types of verification and validation activities. Such continuous checks and assessments of whether the (intermediate) results meet customer requirements are directly in line with the development requirements of ISO 9001, 8.3.4.

---

[3] For further information, please refer for example to Lewrick et al. (2018), pp. 28–29.

A detailed overview of typical process control mechanisms of development processes—each presented for an agile and classical process-based design—can be found in Fig. 2.3.

Control Mechanism	Definition	Classic Development	Agile Development (Scrum)
Requirements	... originate – in analogy to ISO 9000:2015, 3.6.4 – from interested parties or the organisation itself and could be defined, generally implied or obligatory. Therefore, they do not necessarily have to be available in a documented format.	■ requirements specification ■ functional specification ■ standard conditions and contracts	■ client statements in first customer meeting or Sprint planning ■ client feedback obtained on individual steps of the development process or in Sprint Reviews ■ feedback on prototypes Often, customers are only able to provide guidance or make explicit decisions in the course of the process, when possible results become more specific (e.g. in the form of a prototype). In the end, this actually allows a more reliable production of a satisfying result.
Resource Allocation	... directly influences the quality of the process and the process result. There are no general differences between agile and „classic" processes in terms of the need for the availability of financial resources, infrastructure, management support and human resources. The equipment with human resources has to be fitting concerning quantity and quality (e.g. competencies and experience).	Special feature: **Time** ■ control based on time targets is often necessary to meet essential customer requirements ■ Optimising timing, e.g. cycle times, usually requires the control of individual activities in order to accelerate or stabilise the overall process.	Special feature: **Time** ■ in creative processes, certain techniques or methods are employed in very tightly timed periods ("time-boxing") in order to enable all team members to focus on the task at hand ■ the higher concentration of the participants on the respective activity usually leads to better results ■ this principle is applied, for example, in Sprints or in Design Thinking phases
Rules	... include guideline, process instructions and manuals at different levels of obligation. They specify, which activities must be carried out in processes and, if applicable, how these activities have to be executed.	■ policies & guidelines ■ development instructions ■ procedural requirement, e.g. for performing an FMEA based on pre-defined forms.	■ the Scrum Guide ■ instructions for development processes based on SCRUM ■ an extremely detailed procedural requirement is usually inappropriate. Nevertheless, it is quite common to lay down general rules. These can include, for example, the direction, that for every project a Scrum Master has to be appointed who is responsible for forming the team, that has to include people from all departments concerned.

Fig. 2.3 Practical examples of process control mechanisms (own translation of table presented in Japing and Adam (2021), pp. 4–5)

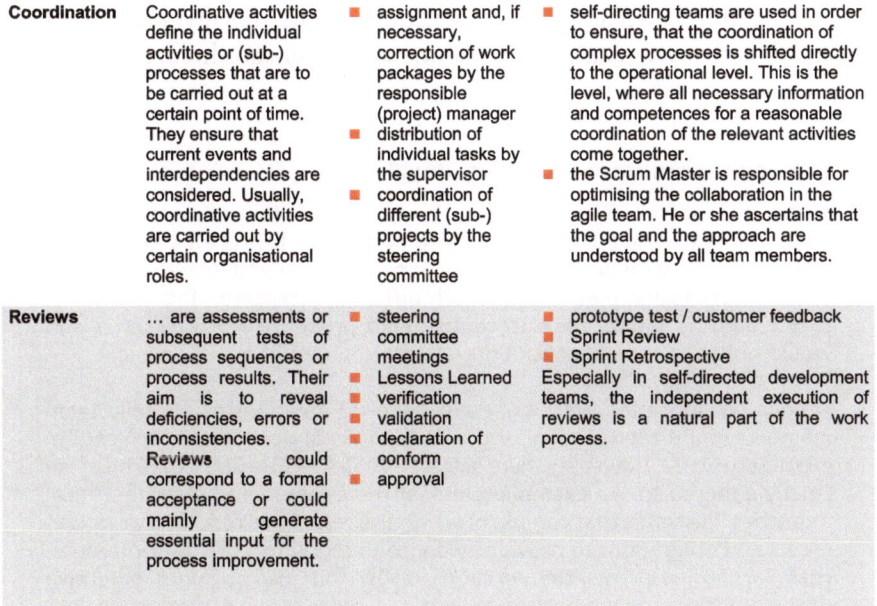

Coordination	Coordinative activities define the individual activities or (sub-) processes that are to be carried out at a certain point of time. They ensure that current events and interdependencies are considered. Usually, coordinative activities are carried out by certain organisational roles.	■ assignment and, if necessary, correction of work packages by the responsible (project) manager ■ distribution of individual tasks by the supervisor ■ coordination of different (sub-) projects by the steering committee	■ self-directing teams are used in order to ensure, that the coordination of complex processes is shifted directly to the operational level. This is the level, where all necessary information and competences for a reasonable coordination of the relevant activities come together. ■ the Scrum Master is responsible for optimising the collaboration in the agile team. He or she ascertains that the goal and the approach are understood by all team members.
Reviews	... are assessments or subsequent tests of process sequences or process results. Their aim is to reveal deficiencies, errors or inconsistencies. Reviews could correspond to a formal acceptance or could mainly generate essential input for the process improvement.	■ steering committee meetings ■ Lessons Learned ■ verification ■ validation ■ declaration of conform ■ approval	■ prototype test / customer feedback ■ Sprint Review ■ Sprint Retrospective Especially in self-directed development teams, the independent execution of reviews is a natural part of the work process.

Fig. 2.3 (continued)

Note: The BIG 5 of process control support all kinds of processes—classic and agile alike.

2.4 Documentation of Agile Practices

Certain documented information is explicitly required by ISO 9001:2015, e.g. a quality policy, quality objectives, or the conformity to design requirements. Of course, it must be possible to present these to the auditor, regardless of whether agile or classical process-oriented practices are used. In most industries, ISO 9001 documentation requirements are only the tip of the documentation iceberg anyway. Most organisations have an almost impossible range of relevant documentation requirements they have to adhere to. For example, they have to provide evidence for being in line with various legal obligations, in accounting as well as otherwise. Nearly all functions in an organisation have their own clearly defined set of documents that is needed to fulfil their tasks or demonstrate results. Finally, there are also documents that are required by customers or other stakeholders.

Example: Freedom for software developers

Especially in software development, the enthusiasm for agile approaches is spreading. Recently a team of software developers told their project manager that they were now an agile development team and would use Scrum. This would of course mean a completely different way of working. Therefore, they would no longer have to comply with predefined quality gates and would not have to worry about documentation requirements anymore.

It's not that simple!

Agile collaboration with e.g. daily stand-up meetings of all agile team members might render some of the usual internal documentation requirements obsolete. However, there are external specifications that must be strictly adhered to. An exemplary look at the banking and pharmaceutical industries illustrates that compliance with the respective regulatory requirements and obligations to provide evidence to regulators are basic prerequisites for business activities. In both cases, this also includes extensive documentation and verification obligations, especially in software development. It is absolutely irrelevant if this is executed by a classic or agile development team. As one interviewee put it, "agile does not mean complete freedom from guidelines or documentation".

The introduction of agility cannot therefore not be used as a reason to dispense with all types of documented information. On the other hand, a culture of a detailed recording of every single consideration and activity would not be compatible with agile practices either. As already mentioned in Sect. 2.2, ISO 9001 does not contain any general requirements on the expected level of detail of documented information. True to the principle of "if it works it works", ISO 9001 requires to ensure that the information is documented in such a way that it can be understood by all people concerned. Depending on the kind of information to be recorded, an instruction in writing might not be an appropriate choice. It is common procedure that some product requirements are documented as sketches or engineering drawings. In private life, the popularity of

YouTube is no coincidence. Learning how to tie a tie or fold an origami paper is far easier when watching the instructor doing it on video. Luckily, according to ISO 9000:2015, section 3.8.6, documented information can take on a wide variety of formats and can be pertained on various media. Thanks to modern media, there is a wealth of possibilities here—videos, podcasts, Lego models, pictures, reports, protocols, graphics, etc. All of these could be used creatively and still in conformity to ISO 9001, as long as the chosen format allows the recipient to process the transmitted information easily and correctly.

Records and Regulations
Analogous to the alternatives to classical recording formats mentioned in the example below, the documentation of rules for the execution of individual activities or processes does not have to be done via the usual guidelines or work instructions. Video tutorials, the recording of virtual seminars for later retrieval, or podcasts are increasingly being found even in classic processes. On the other hand, as already discussed, some agile procedures need clear rules in order to be effective. The incorrect use of Kanban Boards, as already shown in Fig. 1.6, is a prominent example.

However, if, for example, 1-week Sprints are carried out and various creative methods are used in its diverging phase, a detailed documentation of the approaches used is neither necessary nor recommendable. In this case, it would make sense to state only very general key points. Since agile practices are used for real VUCA problems, it is natural that it is not possible to determine beforehand which method is most promising. This is precisely what the team and its facilitator must be able to define and design independently during the course of the project. Relevant for the quality of the process is therefore above all the right team composition and the competent handling of the self-organised work by an experienced facilitator. This will lead eventually to a collectively selected set of agile tools and techniques that allow the team to approach their goal. The arrangements for ensuring these aspects could, if necessary, be set out in an appropriate guideline.

Recording Stand-Up Meetings

It is immediately obvious that it cannot be useful to document a 15-min stand-up meeting in a detailed protocol, which takes a team member 30 min to create. Sometimes, however, essential decisions from the meeting must be recorded and passed on, for example, to team members, customers, or teams working in parallel who were not able to attend the meeting. Depending on the situation, the following alternatives are suitable for documentation purposes:

Boards:
Discussed work packages, upcoming topics, or essential aspects of a decision are written directly on a whiteboard or recorded on post-its by the facilitator during the meeting, which are then attached to a pinboard or flipchart. Such work boards in team rooms are permanently maintained and constitute a constant reference for all members. Absent team members or external parties can be informed by photos of intermediate results, which are saved on the team drive or sent by e-mail. To ensure that everybody understands the board, it is advisable to define a standardised structure (e.g. that of a Kanban board). These boards could also be employed in a virtual version, e.g. Miro boards or Padlets.

Decision photo:
If all team members are present, the wording of a decision made can be briefly recorded on a flipchart. Then all team members stand around the flipchart and briefly hold their thumbs up (or parallel or down). Take a quick cell phone photo of it—done. The photo can be saved, shared using WhatsApp, or sent by e-mail. The image has a unique time stamp, and it is also clear at a glance who was present and how the voting turned out.

Video or audio recording:
During the meeting, the actual approach used is discussed and evaluated in detail. To document why the chosen path was not pursued, either the entire discussion (e.g. for team members who were unable to attend) or the result summarised by the facilitator (e.g. for customers or other teams) is recorded as a video or audio recording. This saves the attendants lengthy explanations in case of later queries. Due to the file size, however, it is better to share these kinds of recordings via a shared drive or Dropbox (sending the file as a link via WeTransfer might also be chosen in case the information is not deemed confidential).

Through a more creative and courageous approach to handling information, new and meaningful forms of documentation can be revealed. These would help to combine ISO 9001 and agile practices with ease.

Modelling of (Agile) Processes

In recent years, many organisations have made considerable efforts to limit the excesses of individually recorded and extremely detailed process modelling. For a medium-sized company, 200 extensively modelled processes in three different levels cost a lot of time and effort, for the yearly updating procedures alone. These efforts are often disproportionate to the actual advantages incurred for process control.

Meanwhile, there are various ideas and tools how the actual procedures used by agile teams could be recorded in process models. Popular solutions are so-called procedural modelling approaches like BPMNEasy. Another solution is provided by subject-oriented modelling as part of a subject-oriented business process management approach. Alternatively, rule-based modelling languages like the Case Management Model and Notation (CMMN), the Declare Framework, or Dynamic Condition Response Graphs are offered.

Some of these approaches attempt to record in detail all actual procedures used in each of the iterative cycles. Subject-oriented modelling, for example, enables each process participant to model his or her process activities individually in the form of subject behaviour diagrams, which should be used afterwards to evaluate and improve the approach (Fleischmann et al., 2012). The technical possibility surely is tempting. However, such a recording hardly makes sense from a management perspective. Individual solutions found for the VUCA world by teams specially selected for a specific problem cannot be transferred to other problems anyway. That is exactly the reason why an agile approach is used. Modelling this in detail does not create any added value.

At the other extreme, agile procedures in process modelling are not even recognisable. In business processes of the first and often also the second level, the sub-processes depicted are naturally so general that agile practices contained therein are not singled out. In some cases, this is done deliberately in order to ensure that a possibly critical auditor is not alarmed by the use of agile practices. From a management perspective, however, this is also not ideal. If agile practices are not isolated but closely intertwined with activities of classical processes, frictions occur. The stumbling blocks mentioned in Chap. 3 are countered more effectively if the resulting friction between agile and classic components is actively

Fig. 2.4   Modelling of agile processes (own illustration)

managed. However, for this to happen, one must be aware of this issue. Therefore, it is recommended to point the presence of agile process components out. For risks that are relevant for the ICS, special notations such as red flags, red triangles, or lightning bolts are used for highlighting the affected process steps. By analogy, special flags could be established for agile practices (see Fig. 2.4).

> **Note: Detailed modelling of agile processes is possible but pointless. However, highlighting (sub-) processes where agile practices are used allows us to actively manage possible frictions.**

References

Adam, P. A. (2021). *Ziemlich beste Freunde - Agilität und ISO 9001*. Artikel, Preprint Version. Hochschule Hannover. Fak. IV, Abt. BWL, Management Series no. 6. Accessed October 14, 2022, from https://doi.org/10.25968/opus-2094

Blatt, M., & Sauvonnet, E. (Eds.). (2017). *Wo ist das Problem? Mit Design Thinking Innovationen entwickeln und umsetzen. 2., komplett überarbeite Auflage*. Verlag Franz Vahlen.

Fleischmann, A., Schmidt, W., Stary, C., Obermeier, S., & Börger, E. (2012). *Subject-oriented business process management*. Springer Berlin Heidelberg.

Japing, T. A. (2018). *Steuerungsmechanismen agiler Prozesse. Masterarbeit*. Hochschule Hannover. Fak. IV, Abteilung BWL, Management series no. 2. Accessed October 14, 2022, from https://doi.org/10.25968/opus-1269

Japing, T. A., & Adam, P. A. (2021). *Die geplante Flexibilität – ISO 9001-konforme Steuerung agiler Prozesse*. Artikel, Preprint Version. Hochschule Hannover. Fak. IV, Abteilung BWL, Management series no. 7. Accessed October 14, 2022, from https://doi.org/10.25968/opus-2095

Lewrick, M., Patrick, L., & Leifer, L. (2018). *The design thinking playbook. Mindful digital transformation of teams, products, services, businesses and ecosystems*. With assistance of Nadia Langensand. Wiley.

DIN EN ISO 9000:2015-11. *Quality management systems – Fundamentals and vocabulary* (ISO 9000:2015); German and English version EN ISO 9000:2015.

DIN EN ISO 9001:2015-11. *Quality management systems – Requirements* (ISO 9001:2015); German and English version EN ISO 9001:2015.

Schwaber, K., & Sutherland, J. (2020). *The scrum guide. The definitive guide to scrum: The rules of the game*. Accessed October 14, 2022, from https://scrumguides.org/docs/scrumguide/v2020/2020-Scrum-Guide-US.pdf#zoom=100

The Agile Alliance. (Ed.) (2001). *Manifesto for Agile Software Development*. Accessed October 14, 2022, from http://agilemanifesto.org/iso/en/manifesto.html

3

Stumbling Blocks of Agile Practices: What to Watch Out for?

Abstract This chapter is dedicated to typical stumbling blocks occurring during the integration of agile practices in everyday business. The design of fitting interfaces to classical (sub-) processes, the selection of suitable agile team members, the handling of redundant managers, and the challenges of retaining common Human Resource Management (HRM) instruments are the biggest obstacles for a successful integration.

Keywords Agility • Agile practices • Agile process • Quality management • Interfaces • Competencies • Managers • Human Resource Management instruments

3.1 Interfaces to the "Normal" World

The view of agility represented here already clarifies in the definition that agility and classical processes must meet each other in an existing organisation. Depending on the contemplated function, well-defined environments and the VUCA world exist in parallel. Now that the zenith of the agility hype has (hopefully) been passed, there are more and more voices amongst the pioneers who no longer focus on a complete "agilisation" of their entire company, but prefer hybrid models (Obmann, 2019, p. 59). This means that every organisation must strive to integrate agile practices and classic processes. Integration that is not or only poorly managed can grow into a considerable risk.

The fundamental question is whether an organisation has only isolated "islands of agility" and otherwise makes use of classical forms of organising its activities. Or whether the proportion of agile practices is so relevant that their processes change their character and become agile. For example, when agile improvement teams in a traditional production environment or in accounting form a kind of quality circle, agility takes place within a strictly limited framework: The core process of the respective department or the organisation as a whole is not affected by these agile practices. Employees spend only a small part of their working time on agile practices and are otherwise assigned to a supervisor. In such cases, an actual integration of the processes is hardly necessary and the organisational challenges are low.

A completely different initial position is given if, for example, in a large development project individual work packages are processed by agile teams, but other packages that are dependent on them are processed in the classical way. If the overall control exerted by the project management follows a standardised methodology and the overall decisions are made by a steering committee staffed with executives, the procedures must be integrated and coordinated. In this case, the essential question is: Who sets the pace? Most of the time, the overall beat is determined by the standard processes. In this case, the agile sub-processes can only determine their workflow freely as long as they stay compatible with the generally specified framework. The balance between the freedom of the agile team, which independently sets goals and defines e.g. Sprint phases, and the control requirements of the overall project requires explicit agreements, a clear commitment, and a lot of trust from all responsible persons. At the same time, such an approach offers the opportunity to put existing regulations to the test and to "clean out" regulations that have long since been dispensable.

Classic and Agile: Programme Management in Development Projects

Start: The Agility Check

The projects and work packages first go through an agility check to determine whether it makes sense to manage them in an agile or classic way. Depending on the result, appropriate procedures apply, which are defined in standard operating procedures (SOPs).

Target Definition via "Target Cloud"

The VUCA world is taken into account by not having to set a complete target from the beginning. Instead, the target is preliminarily defined in the form of a target cloud. This general idea is specified step by step in the upcoming meetings. By that, it will be concretised into a detailed target system that incorporates all learnings from the iterations. In order to be acceptable, the target cloud also has to comply to certain predefined minimum requirements.

(Sub-) Project Management via SOPs

Control structure and methods are individually defined for each project according to task, environment, etc. However, certain aspects are included in the SOPs. For agile projects, the SOPs could contain for example the following aspects:

- A basic project structure as a starting point, from which deviations are allowed if the team decides to do so collectively (this takes account of the necessary flexibility of the process).
- The creation and updating of the target cloud.
- The definition of a fixed project team and the claim that this team will only finish together, i.e. help each other out. The quality rule is: "A work package is only complete when all are finished, i.e. all tests have been successfully completed and the entire documentation is available".
- The definition of responsibilities for certain aspects and decisions (this can also mean "together as an agile team").
- The definition of how transparency about the progress of work is created. This could, but does not necessarily have to be, a Kanban Board.
- Determining when the project objectives are considered as achieved. Depending on the kind of development, this could be a detailed xls table with traceable positive test results or when the prototype is officially accepted by the customer. It should also be clearly defined, when the project is considered completed.
- The need to define associated partner projects and to identify customer contact persons.

(continued)

(continued)

- The documentation of the agreements made with the customers regarding defined aspects, e.g. the determination of the documentation of decisions, the communication processes, and binding technical specifications.
- The execution of daily stand-up meetings in every (sub-) project. During these meetings, essential decisions of the team must be documented, whereby the team itself determines the type of documentation for the duration of the project.

Programme management guidelines
All projects are integrated into higher-level review cycles with binding milestones, at which predefined quality criteria are reviewed and decisions are made on the start of subsequent project processes ("quality gates").

Three quality gates must be passed per project. In agile projects, the first after defining the critical functionalities (usually after the second Sprint), the second after creating the functional prototype, and the third after finalisation for approval by the customer.[1] These quality gates are documented in detail according to a defined structure. It is essential not to set the requirements of the details to be delivered for the first two quality gates too high; otherwise they do not fit to agile projects. "Guard rails" have been developed for this purpose, which must be observed. These requirements can also be adapted individually by agreement, although their permissibility must be checked using a defined risk management check.

In the course of the research project it became clear that in project management the acceptance of rough estimates with iterative improvements over time ultimately leads to more honest planning information and thus to more reliable time and cost planning. Therefore, such an approach provides a basis for faster decision-making. At the same time, more discipline on the part of the deciders is required. With a self-organised team, it is no longer possible to simply enter new requirements "past the process". For example, Sprint Planning is carried out in advance with a very clear definition of the Sprint Goal. Adding new or "forgotten" requirements will lead to an explicit change of the Product Backlog and the final Product Goal,

---

[1] These quality gates are based on the Design Thinking macro-cycle as shown in Lewrick et al. (2018), p. 37.

which is discussed openly and in the case of general acceptance included in the planning. Just because planning becomes more detailed and reliable over time does not mean that change processes could be less strict than in a well-managed, classical "waterfall" project. Every change that affects the Product Goal and Sprint Planning follows a very transparent acceptance process with all agile team members involved.

The transitions from agile to classic practices are particularly important for projects and processes alike. Hence, the conscious design of the communication process including all participants is of special importance. It is usually an advantage if representatives of the departments responsible for subsequent activities are involved in a targeted and regular manner, for example in reviews.

> **Note: Interfaces and transitions between agile and classical process-oriented practices have to be well designed and communicated.**

In addition to the organisational challenges, the introduction of agility also involves mastering challenges of a more individual nature. The different degrees of freedom of agile and classic teams arouse prejudices, envy, and fears. Some people are even overextended with the agile work process.

3.2   Competencies, Readiness, and Self-Organisation

As discussed in Sect. 1.2, the success of agile teams depends largely on the team composition. All relevant perspectives should be integrated and represented by competent team members. Experience with agile work processes or knowledge of agile methods is an advantage. At least some of those involved should have them. It would also be helpful to have an experienced facilitator who is well versed in agile methods to support the selection of methods and ensure their correct application. Many organisations cannot draw — yet — on such a pool of facilitators. So, learning about basic agile principles and methods should be a natural part of any organisation's internal and external advanced vocational training.

Despite offers of agile on- and off-the-job training, not every technically competent employee is willing and able to take responsibility within a self-organising team and to personally deal with agile procedures. Working in an agile team demands a lot from its members. The claim "we are not finished until everyone is finished" stands in direct contrast to a "work to rule" approach. In agile teams, members cannot rely on a working day ending punctually at 5 pm. Time flexibility is just as important as an "agile mindset".

Depending on the publication, different aspects are highlighted in order to define an agile mindset. Basically, the term "mindset" as such describes the logic of thought and action of people or organisations, i.e. the way in which they act or shape the actions of their organisation members and the attitude on which these actions are based (Hofert, 2018, pp. 3–4). From an individual point of view, an agile mindset includes in particular the willingness to change, the ability to self-reflect, the ability to work in a team, and commitment to the common cause. This also includes a certain willingness to take risks, i.e. the courage to try things out, to endure the uncertainty of the chances of success of the chosen path, and to understand mistakes as constructive information on the way to a suitable solution. People who strive for security and prefer routine work naturally feel unhappy and overwhelmed in agile processes. In addition, not everyone is a truly convinced team worker. As one software developer put it: "Not all employees are happy when they have to coordinate their work. Some developers prefer to work alone." According to further statements of the interview partners, which are supported by various sources, a maximum of 50% of all employees[2] really enjoy self-directed agile work. Many publications on agility point out that the younger generation of Gen Y employees — i.e. those born between 1980 and 1999 — per se have a much more flexible mindset and are specifically looking for freedom in their work environment. Unfortunately, this cannot be confirmed in such a general way. In 2014, the FAZ reported that the number of employees in the public sector is rising again and that more and more young people are longing for fixed rules and regulations again (Grossarth et al., 2014).

---

[2] Amongst others Obmann (2019), p. 59, Hofert (2018), p. 4.

The Agility and Innovation Lab

One bank decided to become agile and created an "Agility and Innovation Lab" with rollable tables, writeable walls, and colourful beanbags. People dealing with organisational development projects were expected to work in the lab. In order to set the lab apart from the usual work routines, all people who participated in the lab were expected to come to work in sports shoes and jeans instead of the usual suits. Somehow, it was expected that by changing the work environment and clothing all employees suddenly unfold a miracle of innovation and enthusiasm. Unfortunately, this was not the case. It was still hard to get top management and other employees to accept the solutions worked out in the lab. Some project results were never implemented. Instead, some employees quit and left the bank for a competitor, stating that "they finally want to wear a suit again".

The art is to find the right people. For easing people into agile practices, it is recommended to choose a gentle start. Some organisations start by allowing employees to volunteer for agile (improvement) teams. This means that there is at least a basic willingness for this new way of working. However, if certain expertise is only available in a particular employee, this employee must be recruited for the project. Most interviewees found that it was impossible for them to reliably predict in advance which employees would be able to cope with agile practices and which would not. While some previously critical employees blossomed surprisingly well in an agile team, others, despite initial euphoria, became completely insecure over time and withdrew into themselves. Any organisation that adopts agile practices across the board must be prepared to lose a significant proportion of its employees as a result of this change. This also includes the fact that some agility projects that were started enthusiastically are gradually returning to the status quo, for example by simply delegating decisions from the agile teams back to higher-level managers. The attitude of the managers involved has a significant influence on the members of agile teams. It ultimately determines if they feel comfortable with the agile processes or not.

Flourishing in agile teams

The experiences with members of agile teams vary. One of the interviewees reported a case, where they especially needed to include one specialist in the team. This specialist was an older employee with a very critical attitude. Nobody really dared to address him. However, as they needed his advice, they included him in the team and set out to work with agile procedures. The fascinating experience was that after a short period of time he started to flourish in the team. His negative work attitude stemmed from his frustration concerning the usual routines. For years he provided ideas and advice that was overlooked or ignored by managers and deciders, so that he felt his contributions were not worthwhile. The opportunity to decide together with his team members, act on those decisions, and try things out fulfilled a need that nobody was aware he possessed.

Other interviewees stressed the point that it is sometimes easy to find people that are enthusiastic about working in agile teams but do not have the necessary skills for the tasks at hand. A group that discussed their experiences with team members concluded "learning years aren't earning years — and they aren't agile years, either".

Note: Finding the right people for agile teams is difficult. Easing people into the acceptance of agile practices needs openness to experience, training, and a fitting management attitude.

3.3   Redundant Managers

As explained in Sect. 1.2, an agile team manages itself without an appointed leader or manager. In completely agile processes, middle-level managers are thus redundant. Many managers are not aware of this connection when they make agility the goal of their organisational development. This is why most organisations lack a reasonable clarification of roles.

If agile practices are introduced in a serious manner, the decision-making authority is handed over to the agile team. Previous managers have then two options left (as depicted in Fig. 3.1): "In", i.e. closer to the specialist work and into the control of the process. Or "on", i.e. onto the strategy level, which equals a complete retreat from micro-management

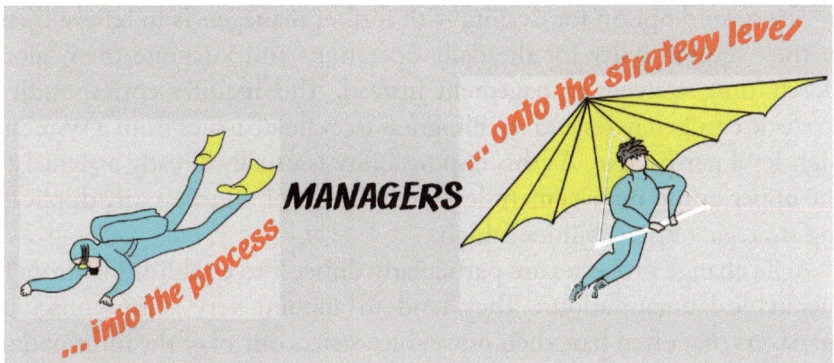

Fig. 3.1 Options for former middle managers (own illustration)

on the operational level. Both options require managers with the appropriate commitment, competencies, and skills.

If an organisation already has a strongly participative leadership culture, it is easier to find managers who are willing to act as facilitators—or "process controllers"—for agile teams. Several of the Scrum Masters interviewed originated from this kind of environment and took over their new role voluntarily, without feeling downgraded by giving up decision-making power. Such a Scrum Master or agile facilitator needs a high degree of communication skills and empathy. Both aspects are connected with good leadership skills anyway. Eventually, the role of these facilitators is to strengthen the self-management skills of the team members. Therefore, an important task is to ensure that organisational strategy and objectives as well as team goals are clearly understood by everyone in the team. Without it, the agile team would not be able to pursue the achievement of these objectives effectively. As one interviewed Scrum Master put it: "I spend a lot of my time now clarifying goals and explaining strategy." Such a facilitator also needs to be ready to defend the team. The facilitator is the one who ensures that the agile team is not disturbed by new demands from the outside, for example during intensive Sprint phases. It is immediately clear that this role cannot be filled by every former manager. As one interviewee put it: "Managers that enjoy power struggles are unsuitable for this."

The second option for dealing with former managers is to relieve them of their responsibility for the daily operations and integrate them more closely into strategic management instead. This requires corresponding strategic competencies and a willingness to evaluate issues from a systemic high-level perspective. As this responsibility is usually already assigned to the upper or top management level, the danger of unnecessarily duplicating strategic responsibilities is high.

Agile change processes are particularly difficult to establish for strongly hierarchical organisations. They tend to appoint very status-conscious managers that often base their power on a strict control of the information flow. In these organisations, a lot of information is only shared with managers and held back from ordinary employees. Managers distribute important aspects on an individual "need to know" basis only. In these environments, agile practices can only be introduced successfully with a clear support from the executive level. Even with a corresponding mandate from top management, this organisation must be prepared to lose many of its current managers who are not able to cope with this agile change. Sometimes, an even worse scenario unfolds, if unwilling people in management positions do not leave the organisation but try to torpedo agile practices instead. Unchanged or missing role descriptions of managers in agile processes offer an ideal basis for this kind of "guerrilla fighter". This kind of managers make life for agile teams difficult in several ways. For example, they can impose detailed reporting requirements for any decisions made by agile teams. Extensive reporting obligations and corresponding critical assessments by these managers very much restrict the self-organisation of their agile teams. Even in case the real influence is low, at least the team members feel controlled and not in charge. In case this manager is still in charge of presenting the team decisions at management meetings, his influence is likely to lead to a change in procedure. Suddenly, the manager's opinion is sought before every decision — for fear of subsequent reprisals in the case of "unacceptable" team decisions. Likewise, this kind of manager might shape a tight error culture by holding the team accountable and publicly exposing them in the event of rejected design prototypes or unachieved (interim) goals. If this manager continues to be responsible for (individual or team) performance appraisals, real agility will quickly lose ground. The only chance against these "guerrilla fighters" as managers is a consistent strengthening of the team's self-organisation by top management. This requires to phase out the former managers from all reporting and decision-making processes and include team representatives (or facilitators) instead. The success of agile practices is highly dependent on strong (top) management support for agile teams, a consistent role concept for managers, and fitting Human Resource Management (HRM) instruments.

Note: Agile teams do not need "classical" managers. Redundant managers can either go into the process or onto the strategic level. In both cases, clear role descriptions and top management support for an agile change are indispensable.

3.4   Human Resource Management Instruments

Many modern concepts of work organisation are currently summarised under the term "New Work", which was originally coined by Frithjof Bergmann. All of these concepts have in common that they are based on the idea of a meaningful form of work in which people experience freedom and self-actualisation.[3] They also include agile practices. This new and free working environment often stands in direct contrast to established HRM instruments. Many HRM tools typically used in organisations were developed on the basis of the classical division of labour stemming from the industrial age. It is therefore not surprising that some of these HRM tools are not suitable for agile practices. Some of them are even counterproductive.

The lynchpin of most target-setting and review procedures in HRM processes is the manager or supervisor. If the latter has disciplinary responsibility, he is personally responsible for career development or—in the case of pay-for-performance systems—for remuneration. Bonuses, commissions, salary increases, and training opportunities are often directly dependent on the performance appraisal by the responsible manager.

---

[3] Details can be found in Bergmann (2019).

If an employee spends his or her working time primarily in an agile team without a manager, the question arises how the performance can be evaluated equitably. As one executive put it, "In self-organised teams, leaders are two steps further back than usual." This means that a manager only sees the results, but not the contribution of the individual team members. Given that different perspectives are a relevant aspect of the team composition, a detailed measurement of individual performance is neither intended nor possible. When an agile team works in a fitting organisational environment, the remaining (top) managers are primarily involved when additional resources are urgently needed or when the strategic direction is unclear. Seldom, managers at higher levels might be called upon as representatives, for example, if despite all efforts a development project finally fails and this must be reported to and discussed with customers, the board of directors, or investors. Former managers only have direct insights into the work process of an agile team if they take on an agile role (e.g. facilitator or Scrum Master, Product Owner, or customer). Even in these cases they can hardly make a reliable assessment.

One possible alternative would be to simply dispense with an appraisal. However, in most organisations, this would also mean to forego the incentives granted through it. It can well be imagined that such an approach would not foster employee commitment in agile teams. Another alternative would be for the team to carry out the appraisal itself. However, all team members would then have to be trained in the predefined proceedings in order to be able to carry out objective assessments of individual contributions. Most probably, the traditional approach (e.g. performance criteria) would also have to be adapted. If only a few parts of the organisation use agile processes, such a revision of a customary framework would be a low priority. The main problem of this alternative constitutes a possible direct competition between team members when resources are scarce, for example, when the team's internal training budget is only sufficient for two or three people. Such a situation naturally influences the result of the assessment.

So far, there is no perfect solution. This might be a reason why many organisations only use agile teams for very special cases and consequently do not adapt their HRM practices. At the same time, more and more organisations that take agility and other New Work practices seriously get rid of their outdated pay-for-performance systems and use a combination of fixed salaries and general bonuses linked to the overall success of the company. Their line of reasoning is as simple as it is compelling: Your agile teams will only be successful if you employ the right kind of people. You have selected the right kind of people when they are dependable team players that are committed to achieving the common goal. In this case, why bother with carrot and sticks?

Note: Pay-for-performance systems are inappropriate for members of agile teams. As an alternative, many organisations use fixed salaries and general bonuses linked to overall success.

Employees in different working environments—shaped by either agile or classic processes—take a critical look at each other. Employees in classic processes quickly feel underprivileged by the supposed freedom of agile colleagues. At the same time, agile colleagues feel discriminated due to the above-mentioned uncertainties regarding performance appraisal. A company that introduces agility on a larger scale must consider which HRM instruments can be used and how. Under these circumstances, a supposedly equal treatment according to the motto "one size fits all" could degenerate quickly into a "one size fits none".

It is precisely the advantage of agile practices, and especially the use of agile methods, that performance can easily be judged by the results obtained. Especially in Scrum projects, the commitment towards the Product and Sprint Goals is detailed out in so-called Increments. These could be understood as parts or modules of the final product. Each Increment must be usable and has to be thoroughly verified. Every work that is accepted as part of an Increment has to meet the Definition of Done, which is formally described and sets the requirements for the quality criteria (Schwaber & Sutherland, 2020, pp. 11–12). This comprehensive approach allows a very detailed assessment of the delivered quality measures, and is often superior to classic development processes.

Such an approach, in turn, corresponds exactly to the overarching principle of ISO 9001:2015: Output matters. In order to assess the reliability of agile processes, the appropriateness of chosen procedures and the fit of the employees involved, goal attainment is a reliable yardstick. In the case of the rescue service, it is the successful emergency care and, in the case of safety goods, the realisation of the customised product within 24 h.

Example: Output measurement in Management Accounting

A few years ago, the leader of the management accounting department of an insurance company realised that he was increasingly asked to provide special reports as input for strategic decisions. He expected this trend of individualised reports to continue, so he invested in the automatisation of processes used for standardised regular reports. The working hours saved enabled his team to react flexibly to unscheduled demands. As already exemplified in Sect. 3.4, they formed individual agile teams for every extraordinary report request. In order to control the output of these teams, they formulated their central goal as follows: "We provide decision-relevant information for all our various customers. This information must be

1. well-founded (which means correct) and
2. be presented at the agreed time (which means adherence to deadlines)."

The goal attainment was consequently measured and the processes were regularly reviewed by the agile teams. In addition, they habitually obtained customer feedback from the initiators of the report requests.

> **Note: Output matters! Judging performance based on goal attainment fits HRM requirements and ISO 9001 alike.**

One critical aspect that has to be mentioned is the fit of working time regulations and agile team processes. According to the ECJ ruling of 14 May 2019, employers in all EU member states are obliged to record the daily working time of their employees objectively and systematically. This serves to ensure that the prescribed rest periods and limits on maximum working hours are actually observed and that workers are properly protected. Depending on the interpretation of these rules, a legally compliant implementation could significantly limit the manageability of self-organised teamwork. For example, the above-mentioned maxim of agile teams "we only go home when everything is ready" could come into conflict with these legal requirements. The extent to which the ECJ verdict limits the freedom of agile work processes in self-organised teams will depend on the detailed regulations for implementation, which are still in the process of being developed. Due to these uncertainties, some organisations started to employ mainly freelancers in agile teams. This is not a reasonable solution. On the one hand, it limits the (internal) experience and skills that can be brought to the team. On the other hand, it could cause frustration on the side of the employees, especially if the opportunity to work in agile teams is seen as particularly attractive or known to enhance the career.

It should also be noted that the employment of freelancers in agile teams can bear the risk of disguised employment. This is especially relevant if those freelancers are completely integrated, for example as part of an existing team, and the work results are not contractually agreed in advance (Kalbfus, 2019). Both aspects are typically given in agile work environments.

Notwithstanding these legal uncertainties, organisations in the VUCA world need to incorporate experience with agility and agile roles into strategic recruitment and career planning. Last but not least, the postulates of Lean Management must also be critically evaluated. Anyone who is serious about agility must be able to put together optimal teams. An organisation that is so "lean" that already the illness of one employee challenges the functioning of essential processes will never be able to accomplish this. For the professional handling of agility, a comparatively generous staffing level based on the existing standard processes is an essential success factor.

References

Bergmann, F. (2019). *New work, new culture. Work we want and a culture that strengthens us.* Zero Books.

Grossarth, J., Löhr, J., & Pennekamp, J. (2014, May 7). Im Sturm und Drang auf den Beamtensessel. In *Frankfurter Allgemeine Zeitung*. Accessed August 23, 2019, from https://www.faz.net/aktuell/wirtschaft/junge-studenten-ziehen-eine-stelle-beim-staat-der-freien-wirtschaft-vor-13028053.html

Hofert, S. (2018). *Das agile Mindset. Mitarbeiter entwickeln, Zukunft der Arbeit gestalten.* Springer Fachmedien Wiesbaden.

Kalbfus, M. (2019, May 29). Zwischen agil und illegal. Neu Arbeitsformen und Scheinselbständigkeit. In *Frankfurter Allgemeine Zeitung* (Nr. 124, p. 16).

Lewrick, M., Patrick, L., & Leifer, L. (2018). *The design thinking playbook. Mindful digital transformation of teams, products, services, businesses and ecosystems.* With assistance of Nadia Langensand. Wiley.

Obmann, C. (2019, June 21). Einer muss den Job ja machen. In *Handelsblatt* (Wochenende 21./22./23. Juni 2019, Nr. 117, pp. 58–59).

Schwaber, K., & Sutherland, J. (2020). *The scrum guide. The definitive guide to scrum: The rules of the game.* Accessed October 14, 2022, from https://scrumguides.org/docs/scrumguide/v2020/2020-Scrum-Guide-US.pdf#zoom=100

4

Outlook: Agile as the New Normal?

Abstract This chapter deals with the rise of agile practices in the disruptive environment of the "new normal". It explains why a broad introduction of home office forces organisations into agility. In this situation, the use of agile tools and a reliable output measurement provides much needed support. The chapter closes with an outlook of the introduction of the concept of agility into the EFQM model as well as into ISO management standards.

Keywords Agility • Agile practices • ISO 9001 • Quality management • Interfaces • VUCA • BANI • Home office • Output • EFQM Model • New normal

4.1 BANI and the Compulsory Agility in Home Office

Due to the Corona pandemic, many organisations were forced to accept home office as the only chance to keep their core processes running. Previously, the belief that the time in home office has to be contained as people are not able to achieve their results without management control was widespread. Some managers seemed to associate home office with the idea of their team members lying relaxed in a hammock, sipping cocktails, and occasionally looking at their computers. Suddenly, the Corona pandemic turned out to become a "boot camp" for the VUCA world. Uncertainty was everywhere, as weekly or daily changes in regulations forced organisations to adapt.

The disruptive environment was so prevalent that VUCA was deemed being an insufficient descriptor for the turmoil that emerged. Thus, the new chaotic world with its unprecedented speed of change was described with a new acronym: BANI. It stands for **B**rittle, **A**nxious, **N**on-linear, and **I**ncomprehensible (Cascio, 2020). From a management perspective, it is absolutely irrelevant if the uncertainty faced is named VUCA or BANI. The important thing to understand is simply that it is impossible to predict even short periods of time and that standardised solutions become more and more obsolete. Thus, agile practices are on the rise.

In many organisations, it was amazing how fast employees adapted to the new situation in home office — despite a lack of workspace, children at home, and Wi-Fi problems. Unsurprisingly, it was harder for managers to accept the new situation, especially in organisational settings, where managers individually assigned most of the activities and reserved their right to make decisions. With limited access to their people in home office, this was no longer a good option. As depicted in Fig. 4.1, it is hardly possible to control activities when the manager sits alone in front of his computer, his people are all in home office, and the technology is on strike. Soon, many organisations realised that in order to ensure smooth operations, frontline staff needed current information right away without a detour via managers. Information processes changed as less management meetings were held and more e-mails were sent directly to the employees. Some managers adapted and took over a facilitating role, sometimes without being aware of it. This often started in virtual meetings, where teams due to the lack of an established netiquette were in dire need of someone facilitating the exchange process.

Looking at the work processes of virtual teams, it has to be stated: A comprehensive introduction of home office forces organisations into agility. A successful convergence towards new ways of workings typically

Fig. 4.1 Management control in home office? (own illustration)

takes place in iterative cycles, or "trial and error". Again, this proved to work better in such an uncertain environment than extensive planning.

Example: Successful communication in home office

In many organisations, teams and their supervisors tried out different varieties of communication processes during the compulsory home office period. At the beginning, some supervisors talked daily to every single one of their team members—via virtual meeting or on the phone. Then they realised that they needed to share more information across the board. There were so many new regulations issued and such a huge amount of news, e.g. from customers, was obtained by different people that access to files was not enough for managing the relevant knowledge in the teams. So, they got everyone into virtual meetings. Because information needed to be exchanged quickly, it was decided to meet daily. They opted for a daily video morning meeting where the current status was discussed for 15 min every morning—results-oriented and with a video camera on. This also ensured that everyone kept regular hours and fell into a reliable routine. As one supervisor of IT developers explained: "with this, we also ensured that some of our colleagues were not remaining in their pyjamas the whole day." After several months, the new processes become more reliable. Therefore, the meeting cycles were adjusted, as every 2 days was enough. In addition, they introduced every Friday afternoon a kind of afterwork meeting, where everybody could join in a relaxed atmosphere and exchange personal topics.

Unfortunately, people are mostly not aware that they crossed the threshold to agility. This is a pity, though, because the possibilities of employing agile procedures and methods for supporting the teamwork are often overlooked. For example, a Kanban Board in its virtual format—as already introduced in Sect. 1.4—would be a very helpful tool for keeping an overview of the assignments currently worked on.

Note: Virtual teamwork is compulsorily agile. Consequently, cooperation could be enhanced by using agile methods and tools.

4.2   Virtual Work: Output Matters

Lately, in an audit, an executive of a public service provider remarked: "I don't believe that all our people really work at home." Such a statement clearly shows how problematic it seems to deal with agile teams from a classical management perspective. There are still a lot of managers around who only feel comfortable when they personally see their people sitting in front of their PCs. However, is this really a reliable sign that their people do a good job? As explained extensively in Sect. 3.4, the performance of agile teams can best be assured through measuring results. This is also true for all kinds of organisations that have widened their home office options. Output matters — also for ISO 9001. In order to achieve this, it becomes clear that in many organisations main prerequisites for a reliable performance management are not yet introduced: formulating clear goals and making results measurable. If, for example, the internal customer of the already introduced management accounting department is provided with a perfectly designed decision-making template, it should not matter whether the brilliant idea was developed in the Design and Agility Lab, a classical meeting room or while swinging in a hammock.

Example: Corona, home office, and management control

Many pandemic plans during Corona relied on safeguarding against staff shortfalls in critical functions. In the IT department of a system-relevant company, for example, they newly created two strictly separated teams. Each team had at least one expert for every main function on board, so that one team was enough to keep the IT running. One week, the head of department and his team were allowed to be in the office. The next week, her deputy and her team were allowed. The respective other team had to stay in home office. For months, members of the different teams were never allowed to see each other—neither at work nor in private. This ensured that in case of one employee spreading his or her Corona infection to all fellow workers, there were still enough experts left to keep the IT running. The approach was required and approved by the responsible regulators. During several months, the IT was kept running despite several Corona incidences in some of the teams. They even managed to successfully finish several development projects. Despite these results, there was more and more pressure exerted by the managing director on the department heads to come to the office for collective meetings. The director deemed his loss of control so critical that he was willing to violate the agreements with the regulators. A typical sign of a manager not being able to tolerate any kind of self-organisation of his teams. Especially during a pandemic, such a critical attitude towards virtual agile practices can produce unnecessary risks.

It has to be stressed again that not all people are willing and able to become part of an agile team and that this is not dependent on age (compare Sect. 3.2). The experiences made in universities during the Corona pandemic support these statements. Many students could neither cope better with technology nor with the virtual world. There seemed to be a huge difference between relaxing in front of the tablet watching Netflix and spending 1 h concentrating on lectures and meetings in front of the screen. According to a scientific study concerning students in a Master programme in Hannover/Germany, the students did all kinds of other things while they supposedly followed the online lectures, namely: cleaned windows, cooked, played Tetris, watched movies, had a look at Amazon, ate breakfast, and hung the laundry (Holtermann, 2020, p. 33). Consequently, there is reason for distrusting some people with regard to work enthusiasm in a virtual environment. Clear goals and measurable results for the self-organised teams are the only way out.

Note: A trusting attitude towards people working in home office should be accompanied by agreeing on clear, measurable goals for the whole team.

4.3   Agility Becomes Respectable

The agility hype nears its end. This does not mean that agility is no longer on the agenda and can be forgotten. On the contrary, especially in the last few years (partly thanks to the Corona pandemic) agile practices became a part of the "new normal". It is now widely accepted that agility is a natural part of an organisation's setup and is indispensable for mastering the challenges of the future. The 15th State of Agile Report shows that agility takes roots on many levels. One aspect is a dramatic increase of the tendency to work remotely, with 56% of the respondents favouring a hybrid approach and 25% even remaining fully remote. The observed rise in the adoption of agile practices in IT and non-IT areas is equally stunning. Reportedly, 86% of software development teams adopted agile practices, but also 29% in operations and even 17% in HRM. The success of these teams is mainly measured by business value delivered (according to 49% of the responses) or by customer/user satisfaction (also 49%). However, one of the main challenges stays the organisational culture which often is at odds with agile values (43% of respondents), although this aspect is already vastly improved in comparison to a few years ago (Digital.ai, 2021).

In the recent version of the EFQM Model, agility is defined in the glossary as "the organisation's ability to change direction/focus in response to an emerging opportunity or threat in a timely way". This definition corresponds to the general view of "somewhat flexible in the VUCA world" and does not make allowance for the equivalent changes in the organisation's setup. However, the use of agility in the EFQM model clearly shows the need for including this special kind of flexibility in every management system that strives for excellence. The criterion part 5.2 of the EFQM Model reads "transform the organisation for the future" (European Foundation for Quality Management, 2021, p. 29) and lists the following practice of an outstanding organisation: It "establishes and utilises agile working approaches, at the same time providing the necessary stability to manage current operations" (European Foundation for Quality Management, 2021, p. 29). Thus, balancing out the need for agility with the need for process orientation in organisational development constitutes a prerequisite for any successful organisation. According to an assessment based on the EFQM data base, organisations perform well if they exhibit "an ability to anticipate future change and … [are] agile and adaptive with their people, processes and resources" (European Foundation for Quality Management, 2021, p. 49). This view is clearly in line with the definitions, distinctions, and statements made here in previous chapters.

The exploration of agility and how it affects quality management was already taken up by ISO/TC 176. Many of the definitions and distinctions made in this book were introduced by the German member of the assigned working group and were finally integrated into the resulting "emerging themes" document. The document finished with the recommendation to further consider and reflect agility in ISO 9001 and related management standards. Agility is seen as an important part of current business realities that will stay relevant in the future. Including it in management standards will ensure that these stay up to date with current organisational practices and continue to assist organisations of all industries in promoting state-of-the-art techniques for organisational development.

Note: Agility is here to stay!

References

Cascio, J. (2020). *Facing the age of chaos.* Accessed September 9, 2022, from https://medium.com/@cascio/facing-the-age-of-chaos-b00687b1f51d

Digital.ai. (2021). *15th State of Agile Report. Agile adoption accelerates across the enterprise.* Accessed September 22, 2022, from https://info.digital.ai/rs/981-LQX-968/images/SOA15.pdf

European Foundation for Quality Management (EFQM). (2021). *The EFQM Model* (Revised 2nd ed.).

Holtermann, S. (2020). *Teleteaching – Provisorium oder langfristige Ergänzung für die Lehre?* Hochschule Hannover. Fak. IV, Abt. BWL, Management series no. 3. Accessed October 14, 2022, from https://doi.org/10.25968/opus-1779

What You Can Take Away from This Book

- Agility is considered a panacea against the VUCA world.
- Agile practices solve VUCA problems and are characterised by self-directed, agile teams and the use of agile, iterative procedures.
- Agile processes are naturally dealing with uncertainty and constant change instead of relying on controlled conditions. They use agile practices to a relevant extent.
- Every organisation can freely combine the design principles process, project and agile. In certified QM systems a minimum of process orientation and agility is indispensable. Every practical approach has to explore the acceptable grey areas.
- Agile teams are no problem for ISO 9001!
- ISO 9001 requires that planning, controlling and monitoring activities are carried out reliably. In a VUCA environment, this can be ensured through agile practices.
- Agile methods are especially rule-based and thus fit very well into an ISO 9001-based management system. This is exactly, why they are so popular.

© The Author(s), under exclusive license to Springer Nature Switzerland AG 2023
P. A. Adam, *Agile in ISO 9001*, Business Guides on the Go,
https://doi.org/10.1007/978-3-031-23588-7

- The BIG FIVE of process control support all kinds of processes—classic and agile alike.
- Detailed modelling of agile processes is possible but pointless. However, highlighting (sub-) processes where agile practices are used allows to actively manage possible frictions.
- Interfaces and transitions between agile and classical process-oriented practices have to be well designed and communicated.
- Finding the right people for agile teams is difficult. Easing people into the acceptance of agile practices needs openness to experience, training and a fitting management attitude.
- Agile teams do not need "classical" managers. Redundant managers can either go into the process or onto the strategic level. In both cases, clear role descriptions and top management support for an agile change is indispensable.
- Pay-for-performance systems are inappropriate for members of agile teams. As an alternative, many organisations use fixed salaries and general bonuses linked to overall success.
- Output matters! Judging performance based on goal attainment fits HRM requirements and ISO 9001 alike.
- Virtual teamwork is compulsorily agile. Consequently, cooperation could be enhanced by using agile methods and tools.
- A trusting attitude towards people working in home office should be accompanied by agreeing on clear, measurable goals for the whole team.
- Agility is here to stay!